Learning Vedic Mathematics on First Principles

Published by :
Lotus Press Publishers & Distributors

Learning Vedic Mathematics on First Principles

Dr. S.K. Kapoor

Lotus PRESS

4735/22, Prakash Deep Building
Ansari Road, Darya Ganj,
New Delhi - 110002

Lotus Press : Publishers & Distributors
Unit No. 220, 2nd Floor, 4735/22, Prakash Deep Building,
Ansari Road, Darya Ganj, New Delhi- 110002
Ph.: 23280047, 98118-38000
• E-mail : lotuspress1984@gmail.com
www.lotuspress.co.in

Learning Vedic Mathematics on First Principles

ISBN: 81-8382-044-1

Printed & Published by : **Lotus Press Publisher & Distributors,** New Delhi-02

INNER VOICE FOREWORD

— FROM A SAINT

Hear your inner voice and be free from the outside limitations. Within is the seat of the Lord and it is here where we have to be face to face with our selves. This origin source reservoir is to be tapped, and this appears to be the urge and intensity of present attempt to reach at 'Vedic mathematics basics' in five steps. These five steps would prepare for the phase and stage to be in communication with the SELF.

The SELF is TRANSCENDENTAL. The seat of the SELF within *Triloki* is the Creator's space. Self established at the seat of Creator's space ensures carrying the Being along the transcendental carriers. It is the bliss of this ensured supply of transcendental carriers which intensifies the urge of the transcending mind to glimpse the inner folds of the transcendental worlds manifesting within Creator's space.

The whole efforts and beginning for it, as such is to be to attain transition from Triloki manifesting as outside mundane macro state world as first state of consciousness for the intelligence to the next transcendental micro state world of the intelligence lively within Creator's space as origin source reservoir of Triloki. Our ghan / body cube, as such, is to be chased for its origin source reservoir within it lively as Chatar mukhi / four head Creator.

Really blissful is the learning path of this chase range. It is transcendentally blessed. Let sadkhas permit their transcending mind to fulfill with bliss of the transcendental worlds / Panch Mukhi / five head lord sustaining the Creator's space.

Lively within cavity of heart:
25-11-2005

In search of ***JYOTI***
manifesting as **NAD**

■

PREFACE

Urge to know ancient wisdom expressed as Vedic knowledge inherently imples acceptance for chase of it the way these systems permit. This as such would mean that the sadkhas for the chase of Vedic systems have to proceed along the established paths of *Sankhiya Nishshta* and *Yoga nishta* which not only run parallel to each but are also complementary and supplementary of each other.

The sankhiya nishtha, on its ultimate analysis amounts to availing the artifices of numbers, for coordination arrangement of organisation format, while on the other hand, the Yoga nishta avails the dimensional formats, for regulation of transcendence and ascendance of mind through the manifested creations. The **TRANS**, also means, the number ahead, and it is this way, the transcending mind ultimately avails the artifices of numbers to sequentially glimpse the ultimate core of the transcendental worlds.

The par-excellence feature of Vedic systems is that these approach the whole range of knowledge as a single Discipline, and as such, on ultimate analysis the organisation of Discipline of knowledge avails the dimensional frames parallel to the artifices of numbers. This, this way, shall be taking us to the artifices of numbers, and the dimensional frames as the source processing tools in terms of which even the self referral feature of transcending mind chasing its own transcending path can also be worked out intellectually.

It is this feature which deserves to be chased for the basics, and here in the present attempt, the urge being to learn and teach Vedic mathematics on its basics, as such, at this phase and stage of exposing ycung minds during Middle school learning, The focus is to be upon the centre of the cube / origin of 3-space being of the format of hyper cube-4 and as such seat of creator's space.

This focus naturally aims to eye the goals of attainments for the young minds for perfection of their intelligence by having transition from the set up of the Cube to transcend ahead for glimpsing the set up of hyper cube-4 along with its transcendental base of the format of hyper cube-5.

Though, the learning all about hyper cubes 4 and 5 is to be the content focus of high school classes, and ahead there to would be the subject focus of the formats of hyper cubes-6 and 7 during higher secondary classes, but the initiations for the foundations of appropriate mental state for those formats are to be ensured during schooling of middle school stage.

This way the learning range of this phase and stage of schooling during classes 5 to 8 is to be centred around the set up of the cube with focus upon its centre /

origin of 3-space as seat of creator's space (4-space) as of format of hyper cube -4. Accordingly the teaching and learning of this range of schooling is phased as five books namely, book-1The teaching of Vedic mathematics, book-2 Learning Vedic mathematics on first principles, book-3 Vedic mathematics basics, book-4 Vedic mathematics skills and book-5 Vedic geometry course, for perfection of intelligence.

These books, in their integrated approach of chase of the range for set up of the cube / 3-space with focus upon its center / origin of 3-space, as source reservoir for the features of the set up, is meant for exposure of young minds to acquire perfection of intelligence while learning basis of basics on first principles along the concrete format of a cube.

The first principles for approaching basis of basics of Vedic mathematics on solid formats are being supplied by the Ganita Sutras, particularly as the ordering principle of Ganita Sutra-1 which makes it as the source Sutra, and as such 'cube' and 'Ganita Sutra-1' are to ever remain within the eye sight of the learners of this phase and stage of schooling.

The aim and goal of learning being the chase of basis of basics on first principles to ensure perfection of intelligence for the young minds, as such heavy responsibiliy lies upon the Vedic mathematics teachers and parents to make the learning really blissful throughout the duration of chase.

The whole range of enlighened souls begining with Maharishi Valmiki and Maharishi Ved Vyas to Maharishi Dayanand and Swami Bharti Krsna TirthaJi Maharaj to Shri Pad Baba and Maharishi Mahesh Yogi, this learning path is being made blissfully lively, and the enlightened teachers are to continue the linege.

With these inspired background, present humble attempt has been made to reach at these books and it is hoped that these may provide required exposure for the young minds with the help of the enlightened teachers, who shall be making their instructions free of technicalities and the knowlede content to flow blissfully within the head s and hearts of the students.

Mr. Kenneth R. Williams and Mr. R. P. Cosmic Kapoor deserve my heartful thanks for remaining intellectually with me during these days of my reaching at these books. Mr. Rohtash Kapoor deserves full appreciation for conducting Vedic geometry course.

Mr. Deepak Girdhar deserves all appreciation for his nice graphics and the computer work.

Credit for nice publication goes to Lotus Press.

The inner voice forward from a saint gives me the satisfaction and confidence as that these efforts may be fruitful for the cause of basics to be brought home to the young minds.

Vedic home,
Rohini, Delhi
25-11-2005

Author

■

CONTENTS

■

SECTION - I
SCIENCE OF LEARNING

1

MIND AND INTELLIGENCE FIELD

1. The 'science of learning' straight a way take us to 'mind and its intelligence field'.
2. At first glance, the 'mind' is the leader of the 'senses'.
3. Mind, as leader of senses, presides over the 'sensory field'.
4. The 'sensory field' has inputs from five 'senses streams'.
5. All what one sees, hears, smells, tastes and feels by touching, in flows into the 'sensory field'.
6. This, 'sensory field', is first state manifestation fold 'presided by the mind'.
7. 'The mind' transcends through this 'manifestation state' and in the process 'creates its next state intelligence field'.
8. It is this feature of the mind transcending through 'the manifestation state and unfolding an intelligence field in the process. which deserves to be chased.

9. The learning essentially is technique of the chase of the transcending mind.
10. The 'sensory field', as such, may be accepted as the waking state manifestation pool of features comprehended with interaction of senses with the manifested world around.
11. The manifested world around, may be designated as Vishwa / Jagat / 3-space / a set up of a cube being representative regular body of 3-space.
12. The learning, as such, takes through the set up of a cube and brings within comprehension and glimpsing of Reality ahead transcending 3-space.
13. The Reality transcending ahead of 3-space, may be designated as a set up of Creator's space / 4-space / hyper cube 4 (domain).
14. This attainment, may be taken as transition and transformation from the 'sensory field' to the 'intelligence field'.
15. In terms of the geometric formats, this range may be approached as beginning with the set up of a cube and reaching the set up of hyper cube 4.
16. The learning process, as such of this range, would be of steps to fully know the set up of a cube and also to know how to transcend there from and attain a transition for the Reality ahead.
17. Therefore, the beginning phase and stage is to have at the starting point a 3-space set up enveloped within a synthetic boundary stitched in terms of eight corner points, twelve edges and six surfaces.
18. Therefore the first focus of learning is to be about

this geometric envelope of the cube manifesting by availing the said 26 geometric entities (8 corner points, 12 edges and 6 surfaces).

19. As such, these 26 geometric entities, become the basic entities, and corresponding to them, would be available the artifices 1 to 26, and these shall be finally taking us to the set of 26 basic elements permitting organization arrangements through the pairing operation for the set of elements / artifices / geometric entities.

20. Knowing this organization arrangements, the learning is to proceed ahead for understanding through comprehension glimpsing of the set up of a cube enveloped within its synthetic boundary.

21. Then it is to reach a phase and stage for the learning process by following the transcending mind, transcending through the set up of the cube and reaching the domain ahead of hyper cube 4, with its domain sustaining the intelligence field created by the transcending mind while being through the set up of the cube as manifested solid boundary of hyper cube 4.

22. This attainment of transcendence through the domain of cube, is to initially take to its centre as origin of 3-space and also as a seat of Creator's space / 4-space.

23. With this the transition process of transcendence through the **linear order** of the set up of 3-space would be set into initiation at the **centre** of the cube.

24. The centre of the cube, as such may be taken as a **sealed origin seat** for the Creator's space / 4-space.

25. The transcending mind while reaching it unlocks the seal at the center / origin and unfolds as an intelligence field manifesting availing 4-space domain.
26. It is this way, the learning process is a process of 'mind creating its intelligence field'. ■

Express and evaluate your comprehension as to:

1. 'Sensory field'
   - -- The extent up till which it may extend.
   - -- Senses reverting back after interaction with boundary of a cube
2. 'Intelligence field'
   - -- Learning process is a process of mind creating its intelligence field
   - -- Transcending mind while reaching center of a cube/origin of 3-space unlocks the seal and enters the domain of creator's space and in the process their emerges an intelligence field.
3. Creator's space / 4-space / Hyper cube-4 / origin of 3-space / centre of a cube

■

2
ATTAINING PERFECTION OF INTELLIGENCE

1. Attainment of 'perfection of intelligence', as practical steps, means, **firstly**, knowing the synthetic boundary of the cube, **secondly** knowing the 'domain of the cube', the way the transcendence through the center of the surfaces, as grid zones, carries solid quantifiers, linear order, and the manifestation of the content as volume matter, and **thirdly** the transcendence through the domain reaching up till the center of the cube and **finally**, the mind unlocking the seal of the center and transcending into the wide open domain of 4-space.
2. This four fold steps of attainment of perfection of intelligence brings within full comprehension and glimpsing of the 3-space set up sustained by the field of intelligence within 4-space domain.
3. The learning process, as such when is guided and supervised by the teacher of the mental state having already glimpsed the Creator's space / 4-space domain / hyper cube 4, is to sequentially result into the requisite learning.

4. The focus as such, for the attainment, is naturally to center the learning process around the set up of the cube, and so centered, the learning process is to sequentially acquire different features of the set up of the cube to be within the comprehension range of one's intelligence.
5. The learning steps beginning with the set up of the synthetic boundary of the cube are to, at the very starting point, is to be of '**pairing process**' for chase of the coordination arrangements of artifices 1 to 26 / geometric entities of 8 corner points, 12 edges and 6 surfaces / 26 basic elements / *chandas Vedanga* (Vedic meters science) / the words formulations of alphabet of 26 letters / the classical and orthodox English words vocabulary.
6. Meters science of Vedas accepts 26 basic meters of four quarter each, and thereby it becomes of the range of 104 akharas / syllables range of measures which measures up till the Vishnu lok / orb of the Sun / 6-space within cuboid / NVF (SUN) =54= NVF (CUBOID).
7. The words formulations of orthodox and classical English of 26 alphabet letters deserve to be chased in the sequential order of the number values generic as formats (NVF) with values for alphabet letters A to Z being 1 to 26 respectively, and the words (formulations) for their number value formats to have summation of the values of the individual letters of the word. Illustratively the WORD as word formulation is to be of number value format of generic range 23 (for W) + 15 (for O) + 18 (for R) + 4 (for D) =60, which is also the generic range of word formulation 'FOUR'.

8. The ***sankhiya nishtha***, which avails artifices of numbers, and ***yoga nishtha*** which avails geometric formats, run parallel to each other, and are also complementary and supplementary of each other, and as such each of it attains the same 'goal'.
9. 'FOUR', as 4 (space), and 'WORD', being of same generic range, shall be making it as that the words formulation are of features of 4-space set up.
10. As such the word formulations of orthodox English, to be taken as, English, deserve to be chased, as such, as formulations of features of 4-space.
11. This way, the learning process, as such is having a very rich source of formulations of features of 4-space, which shall be providing us a best resource for the leads needed at different steps of comprehensions.
12. The start with help which can be had for learning is to understand as that ultimately the whole knowledge gets reduced to the 'counts'.
13. Therefore, counting is the beginning, as well as the end of learning, as well as the learning process of TRANS, the number ahead.
14. This, this way, makes the counting as to be of two orientations, to be designated as of direct (increasing counts) and reverse (decreasing counts) as transcendance and ascendance phenomenon.
15. Therefore, the learning is to be of counting as well as of reverse counting, and parallel to it would emerge the simultaneous addition and subtraction operation maturity as a start with step for attaining

perfection of intelligence, and also as the end attainment of the perfection of intelligence.

16. The number value format, in short NVF, of a given word (formulation) is to be nothing but a counting exercise, which may be had in terms of the following classical words formulations:

    NVF (ENGLISH) = NVF (PAIRING)

    = NVF (POINT)

    = 74

    = NVF (CONE, CONE)

17. As such the first learning step for attaining perfection of intelligence, being taken as the skill of direct and reverse counting, and the second learning step in the direction would be the 'pairing' operation for the artifices 1 to 26; there are precisely 26 primes (including one), up till 100.

18. The above intellectual exercises, on the one hand and the experiential glimpsing at the level of the transcending mind, on the other hand, together are to set the learning process on the path of attaining perfection of intelligence.

19. Real responsibility lies upon teachers (as well as on the parents) to guide and super wise the young minds for their urge and chase for perfection of intelligence.

20. The approach of the learning process for perfection of intelligence being of two fold path of intellectual exercises as well as of experiential exercises, as such it need be ensured that no imbalance stands caused between the pair of folds of learning path.

21. Definitely, it is a very delicate responsibility to

attain balance between the pair of folds of the learning process, as a little carelessness of supervision and guidance is to result into a 'BIOS' for the young minds in favour of one or the other fold of the learning process.

22. Both, the **intellectual fold**, as well as the **glimpsing fold**, are of equal importance and the need for the young minds for ensured timely perfection of intelligence.
23. The sadkhas fulfilled with intensity of urge for attaining perfection of intelligence, shall remain ever respectful to their teachers and parents, and more so with intensity during their guidance and supervision while they being on the learning path.

■

Express and evaluate your comprehension as to:

1. Perfection of intelligence
2. Pairing process
3. Learning process

■

3

SCIENCE OF LEARNING

1. The 'science of learning Vedic mathematics', as such, is a distinct discipline, as it takes care of learning as a system inherently embedded into the creation process itself.
2. Vedic mathematics as a system is of self referral features, as much as that it sequentially unfolds at its every organisation step.
3. This feature when followed chasing the working rules of Ganita Sutras will cover 'science of learning' being its inherent feature because of which it is to ensure the desired degree of perfection of intelligence for the minds being through the processes of the system of Ganita Sutras.
4. As such, the science of learning be taken as running parallel to the sequence and order of the individual Sutras (including Ganita Upsutras.
5. With this, the beginning and starting point of science of learning is to be taken as beginning and starting of the Vedic mathematics system with first Sutra, namely, Ganita Sutra-1 'Ekadhikena

Purvena' / 'one more than before'. And, it to be availed for approaching set up of the cube to initiate the process for perfection of intelligence.

6. Therefore, the comprehension and understanding of Ganita Sutra-1, is to be perfect, as it is in terms of its comprehension and understanding as that the index of attainments of the learning is to be ultimately charted.
7. The 'Ganita Sutra-1' in fact is, like the artifice 1, is a 'full scripture' in itself.
8. The chase of whole range of Ganita Sutra-1 to 16, and also of Ganita Upsutra-1 to 13 may be chased in terms of the functional format of Ganita Sutra-1 itself, as the source sutra of Vedic mathematics system.
9. As such, the **focus** of science of learning, as well as of actual learning, and attainments of learning, **is to be upon** Ganita Sutra-1 as the source ordering principle which is to supply the basics, basis of basics, as well as the first principle for chase of basics on their basis.
10. This chase of basics, basis of basics and first principles in terms of the ordering principle of Ganita Sutras as sequential steps, of 'one more than before', is to, first take through the set up of linear order of 3-space / cube.
11. This attainment of coverage of the domain of 3-space .as linear order, at very next step of manifestation shall be taking from the linear order to the spatial order in terms of the ordering principles of Ganita Sutra-1 as being 'one more than before'.

12. This reaching at spatial order, in fact would amount to having transition from the linear order set up of a 3-space, and same shall be attaining 4-space as of spatial order.
13. This, as such, would mean transcending from 'the sensory domain' as attaining 'the intelligence field'.
14. It is this attainment, inherently imbedded in the process of learning, which is of a central focus. The young minds are to be constantly reminded about this feature of science of learning as a process of learning inherently imbedded into the steps of actual learning in terms of the ordering principle of Ganita Sutras lively within the basics process of ordering while 'counting'.
15. Accordingly, the science of learning which is to formally begin with skill of 'counting', the same also is to end with the values of counts as a process of ordering as a format.
16. So the beginning, as well as the end of science of learning be taken being available at the same 'ment' as that the whole range in between as well silently gets manifested at the same value 'ment'.

■

4

BASIC FORMATS TOOLS

1. The fundamental tool of learning, at this phase and stage, is a 'cube'.
2. The different features of this fundamental tool supply the basic tools of learning.
3. Of these, which helps chase the proportionate symmetry, is the 'mirror'.
4. The **'grids'** are the another basic tools, which with their coordination arrangements of super imposed grids, help chase the transcendence as a flow of streams of content from the centers of the grid zones.

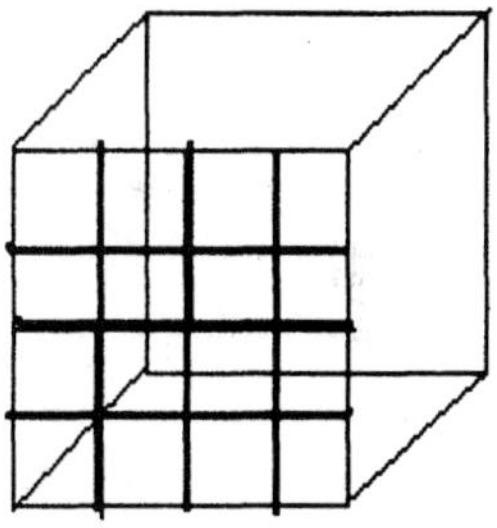

5. Cube as a 'net' of 'grids', is a complete ultimate basic tool of which whole of the cube, as such is nothing but a volume / cave / content lump, and it, in its feature is to be its own 'measure', tool, and the whole set up of a linear order caged within solid grid.

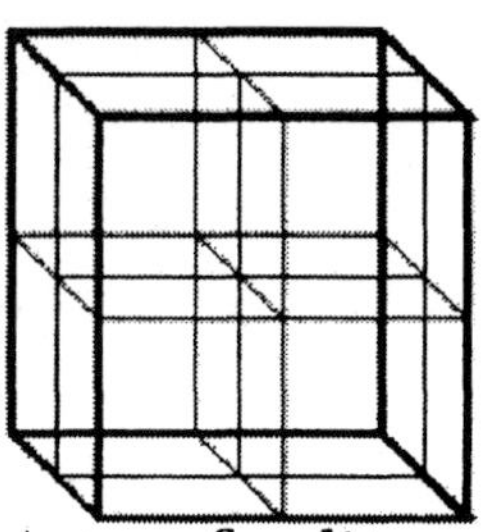

6. Therefore, the linear measuring rod, a pair of cones fixing the center are another set of basic tools.
7. The point / corner point, the interval / edge, square / surface, as geometric entities are another set of basic tools.
8. The pair of prisms is one another set of basic tools.
9. The three dimensional frame, the pair of three dimensional frames of half dimensions and of opposite orientations, constitute another set of basic tools.

First flow orientation

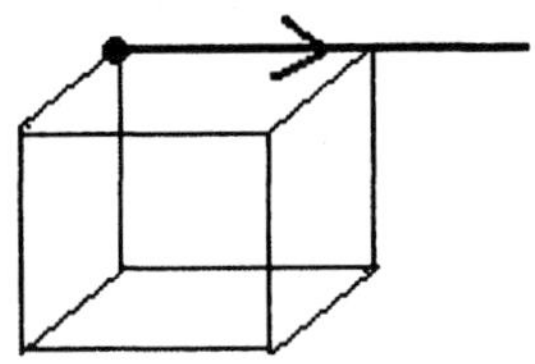

10. The flow as a stream, the flow as a stream splitting into a pair of streams and triple streams constitute another set of basic tools.

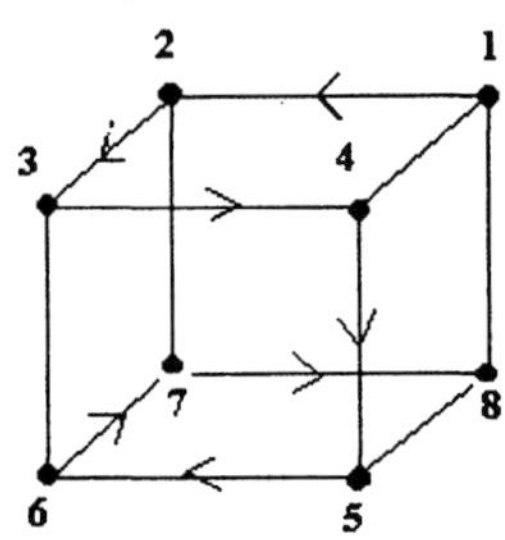

11. Seven edges connecting all the eight corners of a cube, in itself is a class in itself of basic tools of 3-space set up with cube as its representative regular body.

12. The five un manifest edges of the cube as a pair of three edges and two edges groups, is another class of basic tools of 3-space set up with cube as its representative regular body.
13. 1-space in the role of dimension, 2-space in the role of boundary and 3-space in the role of domain of cube constitute a distinct class of basic tools for manifestation of the set up of a cube as representative regular body of 3-space.
14. 4-space in the role of origin manifesting as center of the cube is a distinct class of basic tools of 3-space set up with cube as its representative regular body.

■

5

TRANSITION FROM OLD TO NEW FORMAT

1. The attainments of learning are charted in terms of the potentialities for transition from old to new mental formats for the minds.
2. The mental formats run parallel to consciousness state emerging within the intelligence field.
3. The first consciousness state which emerges within the intelligence field is of **waking state** of consciousness.
4. The waking state of consciousness is of the features of comprehensions of the order of the sensory field capturing the manifestation state of creations.
5. This order of comprehension of the **sensory field** is parallel to the linear order of 3-space with cube as its representative regular body.
6. As such the mental format parallel to it is designated as a **linear order waking state mental format**. It, in short, may be designated as a linear order format. It also may be designated as waking stage mental format. This state also may be

designated as a linear format or a waking state format.

7. When one starts learning with existing mental state, say as of linear format, then this may be taken as the old mental format.
8. When after the learning, transition is attained for the next state of consciousness emerging within the intelligence field, the same is designated as the new mental format.
9. Therefore, the young minds being through this phase and stage of learning are to be taken as there at old mental format being of linear order. When at the end of the course, they shall be acquiring transition for the next state of consciousness, that is of a spatial order 'dream state consciousness' within the Creator's space / 4-space domain / hyper cube 4, it may be said that they have acquired a new mental format.
10. This way, the young minds with old linear mental format shall be acquiring at the end of the course a new spatial order mental format.
11. With old linear order mental format, one was capable of comprehending and understanding only the features of 3-space, while with the new spatial order mental format, one shall be of potentialities to comprehend and glimpse the features of the set up of 4-space / hyper cube 4.
12. Like that, the learning process is of sequential steps which at the end of each phase and stage of learning shall be attaining new mental format in place of the previous stage old mental format.
13. Likewise as at the present phase and stage of

learning which shall be replacing old linear order mental format to new spatial order format, the same, at next phase and stage of learning shall be taking from spatial mental format as old format to solid order mental format of transcendental worlds (5-space) as a new mental format.

14. There are seven states of consciousness, and parallel to it there are seven distinct mental formats which the sadkhas have to be through in their intelligence field. These also may be taken parallel to seven distinct versions of cube parallel to seven geometries of 3-space.

15. This way, heavy responsibility lies upon the teachers to teach and student to learn to attain transition for their existing old mental format order to next new mental format order.

■

6

TRANSITION FROM FORMAT OF CUBE TO THAT OF HYPER CUBE 4

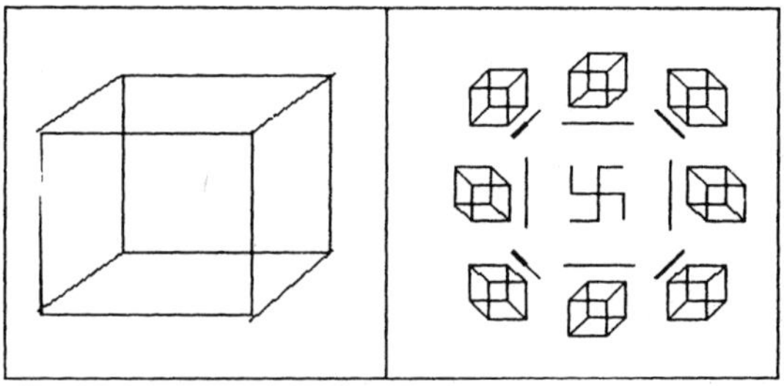

1. The set up of cube as representative regular body of 3-space be taken as the set up of a linear order mental format of waking state of consciousness state as old mental format.
2. The set up of hyper cube 4 as representative regular body of 4-space be taken as the set up of a spatial order mental format of dream state of consciousness state as new mental format.
3. The sadkhas fulfilled with an intensity of urge to fully know and to chase the transition process for its old mental format to that of new mental format, shall first start learning the features of the set up of the cube.

4. One shall chase these features of the set up of the cube at intellectual format as well as at experimental glimpsing state of the transcending mind.

Components of Geometric Envelope		
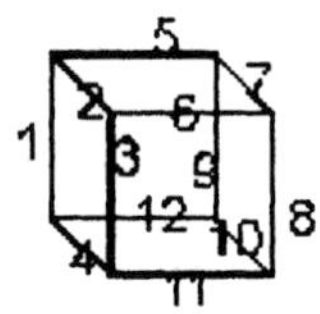	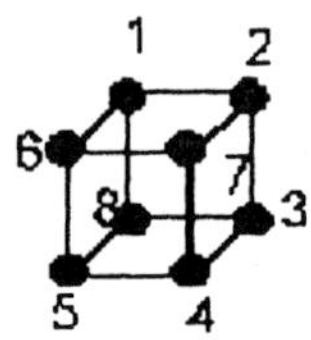	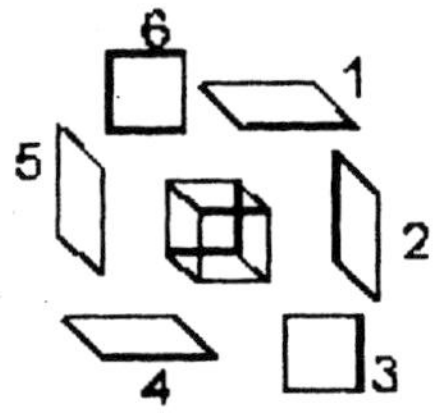

5. One shall transcend through the synthetic boundary of the cube and reach at its domain.
6. One shall further transcend through the domain of the cube and reach at its center.
7. Still further one shall permit the mind to transcend and to un lock the seal at the center of the cube.
8. One shall continue in prolonged sittings of trans till the seal at the center of the cube / origin of 3-space gives way to the transcending mind into the Creator's space / 4-space.
9. One shall continue transcending till one is of the privilege of glimpsing the wide open domain of Creator's space.
10. With center of cube / origin of 3-space giving way to the transcending mind and the wide open domain of Creator's space well coming the glimpsing mind, a process of transition for the old

linear order format of the transcending mind shall take place and in the process the degree of freedom of motion of the cube within Creator's space because of the additional dimension being available, the cube shall be manifesting and additional edge, to be designated as the 13th edge of the cube.

11. The additional edge of the cube shall be availing the additional dimension which would be of a spatial order, and with it, in the process of transcendence and glimpsing of the Creator's space, the transcending mind shall be acquiring transition for its old linear order mental format into a new spatial order mental format parallel to the emergence of spatial format for the additional edge of the cube within hyper cube 4 / 4-space domain / Creator's space.

12. With attainment of a spatial expression format for an additional edge of the cube within hyper cube 4 along its fourth dimension, a process for transition and transformation for the three linear dimensions of a cube / 3-space into three spatial dimensions of hyper cube 4 / 4-space shall be in the process of attainment.

13. It is this attainment of spatial order and also the attainment of an additional dimension which results into the transformation for a 3-space setup into a 4-space set up of four spatial dimensions.

14. It is this set up of four spatial dimension which require a set of eight points for fixation of a point within the hyper cube 4.

15. The center of the cube as a point expression of

hyper cube 4, with opening of its seal, shall be splitting the cube into 8 sub cubes and a 3-space format into 8 octants, and in the process the eight corner points of eight sub cubes of the cube shall be unfolding the seats of eight solid boundary components of hyper cube 4.

16. With this transition for the linear order into a spatial order shall be setting into motion a process of transformation during which the point value hyper cube 4 at center of the cube shall be sequentially evolving into hyper cube 4 of higher values, and ultimately as a full 4-space domain.
17. One shall chase this evolutionary process of the range beginning with zero value hyper cube at center of the cube and at the other end the wide open Creator's space / 4-space domain.
18. One may also intellectually chase the set up of hyper cube 4 as of domain boundary ratio A^4: 8 B^3 as a fourth member of the sequence of hyper cube 4, that is A^N: 2NB $^{N-1}$, N=1, 2, 3, 4 —.
19. Teachers shall also help the students to chase hyper cube 4 as a manifestation of four folds with 2-space in the role of dimension, 3-space in the role of boundary, 4-space as domain and 5-space as origin.

■

7

NECESSITY OF GUIDANCE OF A TEACHER

1. As the old mental format fully grips the whole processing, comprehension, understanding, and it restricts up till the order of the format of mind, as such the knowledge and wisdom of the individual gets caged up till that phase and stage only.
2. One at the mental format of linear order set up of 3-space gets caged within the order and expression of 3-space domain.
3. This way, a big necessity arises for a teacher to guide and supervise to make super-wise the learning process for the student for attaining a transition from its old mental format into the new next phase and stage mental order.
4. The student with old mental format of linear order set up of a cube shall be requiring guidance and help for a learning to have transition and transformation for the old mental format into a new mental format of spatial order of Creator's space / 4-space.
5. It is only with the guidance and supervision of a

teacher that student may expect proper learning for timely attainment of new mental format in place of the old mental format.

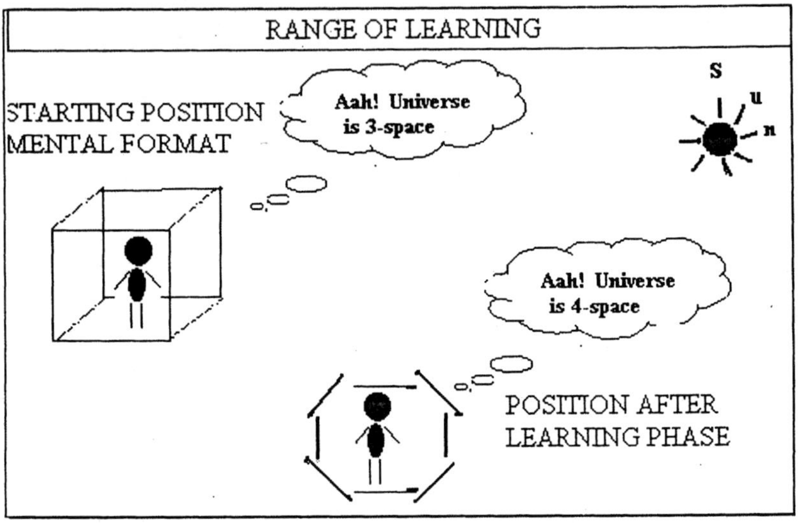

6. No doubt the actual chase exercises at intellectual as well as experimental platform are to be performed by the students for their attainments but the old mental format of the students would be coming in their way, in many ways, and at every such blocking, students shall be enquiring a proper help and guidance to overcome it and as such, for the purpose there is a necessity of learning under the guidance and supervision of the teacher.
7. It is for timely attainments as well that there is a necessity for the guidance and supervision of the teacher.
8. Moreover as this learning process is of two fold path of intellectual chase, as well as, of experiential glimpsing, so without the guidance

and supervision of the teacher, there is every possibility for the imbalance between the two fold of the processing path.

9. The bios in favour of one fold over the other, and the imbalance between the two folds, may result into deviations of the sorts, for whose corrections at later stage may be at the cost of sacrifice of many attainments which simply would not be attainable subsequently because of such earlier stage bios and imbalances.
10. It is the pious duty of the parents to select only the enlightened teachers of proper mental formats for the teaching of their children.
11. It is the solemn duty of the teachers to come true to the entrustment of responsibilities and faith if the parents and innocents souls entrusted for learning under their guidance and supervision. **The students are to be utmost respectful to their learning attainments are to be directly indexed with their intensity of respect for their teachers.**

■

8

LEARNING AND TEACHING ON FIRST PRINCIPLES

1. Learning and teaching on first principles is an interactive process but the fact that the teachers and student are at different mental formats of acquisition and glimpsing, as such heavy responsibility lies upon the teachers to be persuasive during their guidance and supervision to make the process really interactive.
2. The book 'learn and teach Vedic mathematics' of the author published by the same publisher, may be of help.
3. 'Cube' is to be the real measure of learning and teaching on first principles.
4. Hyper cube 4 would be the real measure of attainments of teaching and learning on first principles.
5. One way to look at the process of learning would be to take as that cube is emerging from a zero / void / point and it also fades into a void.
6. Another way to look at the process of learning is that it is the set up of 'one' / unit / line / 1-space

/ linear order which results into the set up of a cube, and this set up can be taken as constituting a line.

Interval	*Square*	*Cube*	*Hypercube-4*
		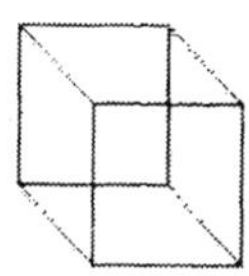	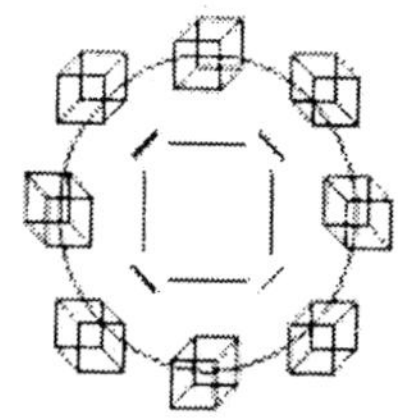

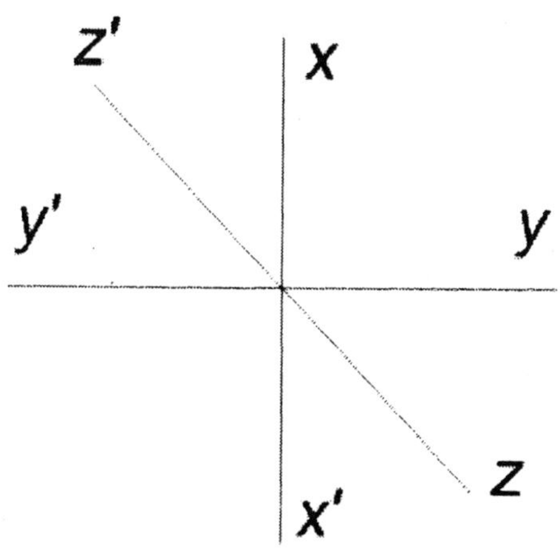

xyz
xyz'
x'yz
x'yz'
x'y'z
x'y'z'
xy'z'
xy'z

7. Still another way to look at the process of learning is that it is the set up of 2 / plane / 2-space which is attaining transition for the linear order of the set up of a cube and this set up also can be taken as enveloping the spatial order Creator's space.
8. Still further the set up of the cube may be captured in a static and dynamic states.

9. Still further the set up of 3-space may be captured in its different roles as a domain / as cube, as a boundary, as solid boundary of Creator's space / 4-space / hyper cube 4.
10. Still further the set up of 3-space may be captured at its point value at origin of 2-space / center of a square.
11. Still further the set up of cube may be captured as a manifestation with 1-space as dimension, 2-space as boundary, 3-space as domain.
12. This way, the teaching and learning of 3-space on first principles may be of many features, and for full comprehension of the set up of 3-space, all these features need be captured, by learning in a geometry class in Ayurveda garden.
13. One outline of these features comprehension may be of following steps.

    Step-1 Introduction

    Teacher-disciple joint prayer (Kathoupanihad prayer) 1.2 Mathematical approach (Yoga Churamanipanishad-71) 1.3 Mathematics of one 1.4 Shad Chakra format 1.5 Vedic mathematics 1.6 Macro World and Micro world 1.7 Macro (physical) world basis of syllabus of class VIII 1.8 Macro (Atomic) world basis of syllabus of class IX 1.9 Nuclear world basis of syllabus of class X 1.10 Supreme order basis of syllabus of class-XI.

    Step-2 Macro (physical) world basis of syllabus of class VIII

    2.1 Physical world 2.2 Macro world versus micro world 2.3. Three space format 2.5 Linear dimensional order 2.5 One space in the role of dimension 2.6 Two space in the role of boundary 2.7 Three space in the role of domain 2.8 Four space in the role of origin. 2.9 Three manifestation layer 2.10 Five space format for transcendence.

Step-3 Three space and three space bodies

3.1 Three space 3.2 Three space content 3.3 Three space as domain 3.4 Three space bodies 3.5 Mathematics of three space 3.6 Euclidean geometry 3.7 Theory of equation 3.8 Solid bodies 3.9 Motion of bodies 3.10 Limits of three space mathematics.

Step-4 Ganita Sutras (Mathematics of the linear order)

4.1 Ganita Sutras 4.2 Ganita Sutra-1 4.3 Ganita Sutra-14 4.4. Mathematics of linear order 4.5 Ganita Sutra-II 4.6 Ten place value system 4.7 Ganita Sutra-III 4.8 Multiplication operation 4.9 Ganita Sutra-IV 4.10 Reflection operation.

Step-5 Properties of the cube

5.1 Cube 5.2 Three dimensional fixation of cube 5.3 Four versions of cube 5.4 Five latent edges of cube 5.5 Six surfaces of cube 5.6 Seven versions of the cube 5.7 Eight octant cut of the cube 5.8 Nine points fixation of cube 5.9 Ten directions of fixations of cubes 5.10 Center of cube.

Step-6 Properties of the sphere

6.1 Sphere 6.2 Three dimensional fixation of sphere 6.3 Surfaces area of sphere 6.4 Volume of sphere 6.5 Polar coordination 6.6 Solid angles 6.7 Circle and sphere 6.8 Plane geometry 6.9 Circular trigonometry 6.10 Spherical trigonometry.

Step-7 Motions

7.1 Motion 7.2 Speed 7.3 Velocity 7.4 Acceleration 7.5 Motion on a line 7.6 Motion in a plane 7.7 Motion in three space 7.8 Concept of time 7.9 Time as fourth dimension 7.10 Time and geometric dimensions

Step-8 Curves and graphs

8.1 Track of moving point in one space 8.2 Track of moving point in two space 8.3 Track of moving point in three space 8.4 Different versions of intervals 8.5 Plane curves 8.6 Families of the plane curve 8.7 Three space curve 8.8 Trajectory 8.9 Graphs 8.10 Statistical graphs

Step-9 Solid bodies

9.1 Cube 9.2 Sphere 9.3 Cylinder 9.4 Cone 9.5 Tetrahedron 9.6 Egg 9.7 Tree 9.8 Bird 9.9 Animal 9.10 Man

Step-10 Structured points

10.1 Point of one space 10.2 Set theory 10.3 Point of plane 10.4 Cartesian geometry 10.5 Point of three space 10.6 Interval as collection of points 10.7 Plane as collection of lines and points 10.8 Solid bodies as collection of planes 10.9 Points of lines have length 10.10 Point of plane have are.

■

SECTION - II

TABLES 100 × 100

1

INTRODUCTORY

Vedic mathematics approach to tables may be designated as 'digit by digit approach', as every digit of the number is managed separately. Illustratively, if one is to write the table of '123', then this assignment is split as 3 distinct exercises, corresponding to the 3 digits of the number '123', namely, '3', '2' and '1'. These 3 digits are taken as first entries of 3 columns as under:

Table Value	Third column	Second column	First column	Entry
1 × 123	1	2	3	First

The Vedic systems workout table of '123' for its second entry, as set of 3 distinct entries for 3 columns as tables of 3, 2 and 1 respectively which together emerge as under:

Table Value	Third column	Second column	First column	Entry
1 × 123	1	2	3	First
2 × 123	2	4	6	Second

In a sequence, the third, fourth and following entries emerge as the flow charts of tables 3, 2, 1 respectively.

Table Value	Third column	Second column	First column	Entry
1 × 123	1	2	3	First
2 × 123	2	4	6	Second
3 × 123	3	6	9	Third
4 × 123	4	8	12	Fourth

Note The fourth entry for fist column as 12 being of double digit, it shall be pushing second digit namely '1' as carry forward for the second column to make it '8+1' =9 and thereby the fourth entry for the table 4 x 123 shall be of value as:

4 × 123	4	9	2	Fourth

The table '123' shall continue unfolding as under:

Table Value	Third column	Second column	First column	Entry
1 × 123	1	2	3	First
2 × 123	2	4	6	Second
3 × 123	3	6	9	Third
4 × 123	4	9	2	Fourth
5 × 123	5	10	5	Fifth

Note : Fifth entry for the second column shall be pushing second digit of '10' that is '1' as carry forward for the first column and it shall be yielding value for first column at the stage as '5 +1 =6'. With it the table shall be of values for these entries as under:

Table Value	Third column	Second column	First column	Entry
1 × 123	1	2	3	First
2 × 123	2	4	6	Second
3 × 123	3	6	9	Third
4 × 123	4	9	2	Fourth
5 × 123	6	0	5	Fifth

These carry forward operations are simple one step mental operations, and as such the whole exercise of table '123' can be mentally worked out as tables of '3', '2' and '1' together with the carry forward operations at appropriate stages.

1.1 Conversion of bigger digits (6, 7, 8, 9)

This digit by digit approach of Vedic systems has an added feature of its working only in terms of smaller digits, viz. (0, 1, 2, 3, 4, 5). The bigger digits namely (6, 7, 8, 9) are suitably converted for their handling in terms of smaller (positive and negative) values of range 1 to 4. This technique of conversion has its base in acceptance of negative numerals (-1, -2, -3, -4, -5, -6, -7, -8, -9) parallel to the positive numbers (1, 2, 3, 4, 5, 6, 7, 8, 9) are also provided special symbols as:

$$(\bar{1}, \bar{2}, \bar{3}, \bar{4}, \bar{5}, \bar{6}, \bar{7}, \bar{8}, \bar{9})$$

The conversion of bigger digits, as such, is attainable in terms of simple arithmetic equations of digits and their deficiencies from the base/place value (10), as under:

$$6=10-4=1\bar{4},\quad 7=10-3=1\bar{3},\quad 8=10-2=1\bar{2},\quad 9=10-1=1\bar{1},$$

These conversions for digits 6, 7, 8 and 9 as:

$$1\bar{4},\ 1\bar{3},\ 1\bar{2},\ 1\bar{1}$$

though are of double digits format, but both digits for each expression here are of values of range 1 to 4. With it the mental operations become simpler as only the values 1 to 4 are required to be handled.

The double digits expression format is parallel to the corresponding double digits of positive values alone. The second digit can be carried forward, and the first digit can be arithmetically operated parallel to the corresponding positive values digits.

As such, as a net result, as is the claim of the Vedic systems that once the child knows the tables up till 5 x 5, the rest whole range up till any length, may it be 100 x 100 or 1000 x 1000 or even larger, the same can be handled and worked out in terms of the tables 5 x 5. In fact, the whole unfolding process, as has been the above case of '123', is to be there, as digit by digit unfolding, which for all practical purposes is to be simply the additions/subtractions of values 1 to 5, and that's all with the technique of carry forward/backward of '1'.

1.2 Conversion table for double digits numbers 00 to 99

Number	01	02	03	04	05	06	07	08	09	10
Conversion	01	02	03	04	05	$1\bar{4}$	$1\bar{3}$	$1\bar{2}$	$1\bar{1}$	10

Number	11	12	13	14	15	16	17	18	19	20
Conversion	11	12	13	14	15	$2\bar{4}$	$2\bar{3}$	$2\bar{2}$	$2\bar{1}$	20

Number	21	22	23	24	25	26	27	28	29	30
Conversion	21	22	23	24	25	$3\bar{4}$	$3\bar{3}$	$3\bar{2}$	$3\bar{1}$	30

Number	31	32	33	34	35	36	37	38	39	40
Conversion	31	32	33	34	35	$4\bar{4}$	$4\bar{3}$	$4\bar{2}$	$4\bar{1}$	40

Number	41	42	43	44	45	46	47	48	49	50
Conversion	41	42	43	44	45	$5\bar{4}$	$5\bar{3}$	$5\bar{2}$	$5\bar{1}$	50

Number	51	52	53	54	55	56	57	58	59	60
Conversion	51	52	53	54	55	$1\bar{4}\bar{4}$	$1\bar{4}\bar{3}$	$1\bar{4}\bar{2}$	$1\bar{4}\bar{1}$	$1\bar{4}0$

Number	61	62	63	64	65	66	67	68	69	70
Conversion	$1\bar{4}1$	$1\bar{4}2$	$1\bar{4}3$	$1\bar{4}4$	$1\bar{4}5$	$1\bar{3}\bar{4}$	$1\bar{3}\bar{3}$	$1\bar{3}\bar{2}$	$1\bar{3}\bar{1}$	$1\bar{3}0$

Number	71	72	73	74	75	76	77	78	79	80
Conversion	$1\bar{3}1$	$1\bar{3}2$	$1\bar{3}3$	$1\bar{3}4$	$1\bar{3}5$	$1\bar{2}\bar{4}$	$1\bar{2}\bar{3}$	$1\bar{2}\bar{2}$	$1\bar{2}\bar{1}$	$1\bar{2}0$

Number	81	82	83	84	85	86	87	88	89	90
Conversion	$1\bar{2}1$	$1\bar{2}2$	$1\bar{2}3$	$1\bar{2}4$	$1\bar{2}5$	$1\bar{1}\bar{4}$	$1\bar{1}\bar{3}$	$1\bar{1}\bar{2}$	$1\bar{1}\bar{1}$	$1\bar{1}0$

Number	91	92	93	94	95	96	97	98	99	100
Conversion	$1\bar{1}1$	$1\bar{1}2$	$1\bar{1}3$	$1\bar{1}4$	$1\bar{1}5$	$10\bar{4}$	$10\bar{3}$	$10\bar{2}$	$10\bar{1}$	100

1.3 Tables 100 × 100

The above conversion table of digits 01 to 99 in terms of digits (positive/negative) of values range 0 to 5 shall be of great help to workout the tables of whole range up till 100 x 100 by additions/subtractions of values (positive/negative) only of the range 0 to 5 together with the simple mental mathematics operation of carry forward/backward of value '1' wherever it emerges or is required as second digit. Here below is taken up the illustrative case of table 99.

1. Conversion of number 99

From the above table it can be seen that 99=101.

2. The three digit expression for the number 99, can also be converted into four digits expression as 0101.
3. The four digits expression can be grouped as a pair of double digits as '01' and '01'.
4. This well indicates that the sequential increase by '1' and sequential decrease '1' is expected by the said pair of parts of four digit expression for 99.
5. The above may be availed for unfolding table 99 as table 0099, along a pair of columns, as under, with first entry for first column below being 99, and the first entry for the second column being '00'.
6. The above expectation of sequential decrease by '1' for the first column and sequential increase by '1' for the second column, shall be accordingly unfolding second entry for both columns accordingly as under following the above increase and decrease rule:

Table Value	Second column	First column	Entry / Rule
1 × 99	00	99	First
	Increase by '1'	Decrease by '1'	Rule
2 × 99	01	98	Second

The subsequent steps to unfold of their own as sequential increase and decrease values of '1' at each step:

Table Value	Second column	First column	Entry / Rule
1 × 99	00	99	First
	Increase by '1'	Decrease by '1'	Rule
2 × 99	01	98	Second
2 × 99	02	97	Third

There are many interesting features of this table which shall be taken up for their focus at appropriate stage.

1.4 Many Alternative Approaches are Available

There are a large number of alternative approaches are available to roll down tables. The basic amongst them is the above digit by digit approach in terms of only smaller digits values. This approach in its generality extend to cover any length of numbers line. It is equally applicable to any number value system, though here the focus only centers around ten place value system.

At the cost of repetition, it is being added because of the importance of the aspect, as that the digit by digit approach is not restricted only to ten place value system and that it is equally applicable to all place value systems and for senior students it would be blissful area to work out tables for different place value systems, illustratively up till the range of 8 x 8 for eight place value system.

The second approach to tables is by internal split of the numbers. Illustratively if one is to work out the table of 36. It can be worked out as 30+05+01, which at second step would yield value being 60 + 10 + 02, as second steps for tables 30, 05, and 01.

The table 30 is essentially a working skill of table of 3. There for the working skills of tables '1', '3' and '5' shall be completing the exercise for the table '36' by following the above split up. Then there would be many alternative splits for 36 and this way many alternatives being available.

FULL TEXTS OF TABLES 99 × 100 and 98 × 100

TABLE 99

1. Re-express 99 in the following format

Second Column	First Column
00	99

2. Write expression for progressions rule as follows

$$00\ 99 = 01\ \overline{01}$$

3. Progressions rules

for first column	decrease by one at each step from step 2 onwards
for second column	increase by one at each step from step 2 onwards

4. Progression from step 1 to step 2

under first column value 99 – 01 = 98

under second column value 00 + 01 =01

5. Chase for the initial steps

Step No.	Second Column	First Column	Table Step	Value of the step
1	00	99	1 × 99	0099
2	00 + 01 = 01	99 – 01 = 98	2 × 99	0198
3	01 + 01 = 02	98 – 01 = 97	3 × 99	0297

Step No.	Second Column	First Column	Table Step	Value of the step
4	02 + 01 = 03	97 – 01 = 96	3 × 99	0396
5	03 + 01 = 04	96 – 01 = 95	4 × 99	0495
-	-	-	-	-

6. Test

   Sum of the values of both the columns = 99

   Illustratively 00 + 99 = 99, 01 + 98 = 99 and so on.

7. Complete table of 99 for all the 100 steps

01 × 99	0099	9900	100 × 99
02 × 99	0198	9801	99 × 99
03 × 99	0297	9702	98 × 99
04 × 99	0396	9603	97 × 99
05 × 99	0495	9504	96 × 99
06 × 99	0594	9405	95 × 99
07 × 99	0693	9306	94 × 99
08 × 99	0792	9207	93 × 99
09 × 99	0891	9108	92 × 99
10 × 99	0990	9009	91 × 99
11 × 99	1089	8910	90 × 99
12 × 99	1188	8811	89 × 99
13 × 99	1287	8712	88 × 99
14 × 99	1386	8613	87 × 99
15 × 99	1485	8514	86 × 99
16 × 99	1584	8415	85 × 99
17 × 99	1683	8316	84 × 99

18 × 99	1782	8217	83 × 99
19 × 99	1881	8118	82 × 99
20 × 99	1980	8019	81 × 99
21 × 99	2079	7920	80 × 99
22 × 99	2178	7821	79 × 99
23 × 99	2277	7722	78 × 99
24 × 99	2376	7623	77 × 99
25 × 99	2475	7524	76 × 99
26 × 99	2574	7425	75 × 99
27 × 99	2673	7326	74 × 99
28 × 99	2772	7227	73 × 99
29 × 99	2871	7128	72 × 99
30 × 99	2970	7029	71 × 99
31 × 99	3069	6930	70 × 99
32 × 99	3168	6831	69 × 99
33 × 99	3267	6732	68 × 99
34 × 99	3366	6633	67 × 99
35 × 99	3465	6534	66 × 99
36 × 99	3564	6435	65 × 99
37 × 99	3663	6336	64 × 99
38 × 99	3762	6237	63 × 99
39 × 99	3861	6138	62 × 99
40 × 99	3960	6039	61 × 99
41 × 99	4059	5940	60 × 99
42 × 99	4158	5841	59 × 99
43 × 99	4257	5742	58 × 99
44 × 99	4356	5643	57 × 99
45 × 99	4455	5544	56 × 99

46 × 99	4554	5445	55 × 99
47 × 99	4653	5346	54 × 99
48 × 99	4752	5247	53 × 99
49 × 99	4851	5148	52 × 99
50 × 99	4950	5049	51 × 99

8. Note. One of the special features of this table is that the values for steps 1 to 50 with the reflection pair values become the steps 100 to 51 in that sequence and order as is evident from the above tabulation.

TABLE 98

1. Re-express 98 in the following format

   Second Column First Column

   00 98

2. Write expression for progressions rule as follows

   00 98 = 01 02

3. Progressions rules

for first column	decrease by two at each step from step 2 onwards
for second column	increase by one at each step from step 2 onwards

4. Progression from step 1 to step 2

   under first column value 98 − 02 = 96

   under second column value 00 + 01 =01

5. Chase for the initial steps

Step No.	Second Column	First Column	Table Step	Value of the step
1	00	98	1 × 98	0098
2	00 + 01 = 01	98 – 02 = 96	2 × 98	0196
3	01 + 01 = 02	96 – 02 = 94	3 × 98	0294
4	02 + 01 = 03	94 – 02 = 92	4 × 99	0392
5	03 + 01 = 04	92 – 02 = 90	5 × 99	0490

6. Test

   Sum of the values of both the columns is decreasing by 1 at each step Illustratively 00 + 98 = 98, 01 + 96 = 97 and so on.

7. Complete table of 98 for all the 100 steps

01 × 98	0098	9800	100 × 98
02 × 98	0196	9702	99 × 98
03 × 98	0294	9604	98 × 98
04 × 98	0392	9506	97 × 98
05 × 98	0490	9408	96 × 98
06 × 98	0588	9310	95 × 98
07 × 98	0686	9212	94 × 98
08 × 98	0784	9114	93 × 98
09 × 98	0882	9016	92 × 98
10 × 98	0980	8918	91 × 98
11 × 98	1078	8820	90 × 98
12 × 98	1176	8722	89 × 98
13 × 98	1274	8624	88 × 98
14 × 98	1372	8526	87 × 98

15 × 98	1470	8428	86 × 98
16 × 98	1568	8330	85 × 98
17 × 98	1666	8232	84 × 98
18 × 98	1764	8134	83 × 98
19 × 98	1862	8036	82 × 98
20 × 98	1960	7938	81 × 98
21 × 98	2058	7840	80 × 98
22 × 98	2156	7742	79 × 98
23 × 98	2254	7644	78 × 98
24 × 98	2352	7546	77 × 98
25 × 98	2450	7448	76 × 98
26 × 98	2548	7350	75 × 98
27 × 98	2646	7252	74 × 98
28 × 98	2744	7154	73 × 98
29 × 98	2842	7056	72 × 98
30 × 98	2940	6958	71 × 98
31 × 98	3038	6860	70 × 98
32 × 98	3136	6762	69 × 98
33 × 98	3234	6664	68 × 98
34 × 98	3332	6566	67 × 98
35 × 98	3430	6468	66 × 98
36 × 98	3528	6370	65 × 98
37 × 98	3626	6272	64 × 98
38 × 98	3724	6174	63 × 98
39 × 98	3822	6076	62 × 98
40 × 98	3920	5978	61 × 98
41 × 98	4018	5880	60 × 98
42 × 98	4116	5782	59 × 98

43 × 98	4214	5684	58 × 98
44 × 98	4312	5586	57 × 98
45 × 98	4410	5488	56 × 98
46 × 98	4508	5390	55 × 98
47 × 98	4606	5292	54 × 98
48 × 98	4704	5194	53 × 98
49 × 98	4802	5096	52 × 98
50 × 98	4900	4998	51 × 98

Sum Up

1. Counting is just a table of '1'.
2. The table of '1' (table '1') is known as counting table.
3. As here we are interested to learn to write tables '1 to 100' of 100 steps each, as such, let us, first of all tabulate here under the counting up till 100, and to designate it as table '1', that is, 1 × 100.

1.	2.	3.	4.	5.	6.	7.	8.	9.	10.
11.	12.	13.	14.	15.	16.	17.	18.	19.	20.
21.	22.	23.	24.	25.	26.	27.	28.	29.	30.
31.	32.	33.	34.	35.	36.	37.	38.	39.	40.
41.	42.	43.	44.	45.	46.	47.	48.	49.	50.
51.	52.	53.	54.	55.	56.	57.	58.	59.	60.
61.	62.	63.	64.	65.	66.	67.	68.	69.	70.
71.	72.	73.	74.	75.	76.	77.	78.	79.	80.
81.	82.	83.	84.	85.	86.	87.	88.	89.	90.
91.	92.	93.	94.	95.	96.	97.	98.	99.	100.

4. Vedic systems avail positive and negative numerals simultaneously to simplify the operations in terms of smaller value numerals, namely, 1, 2, 3, 4 & 5. The bigger numerals namely 6, 7, 8 & 9 are replaced as

$6=1\bar{4}, 7=1\bar{3}, 8=1\bar{2}, 9=1\bar{1}$

5. Availing this technique, the above counting table is rewritten as under

1	2	3	4	5	$1\bar{4}$	$1\bar{3}$	$1\bar{2}$	$1\bar{1}$	10
11	12	13	14	15	$2\bar{4}$	$2\bar{3}$	$2\bar{2}$	$2\bar{1}$	20
21	22	23	24	25	$3\bar{4}$	$3\bar{3}$	$3\bar{2}$	$3\bar{1}$	30
31	32	33	34	35	$4\bar{4}$	$4\bar{3}$	$4\bar{2}$	$4\bar{1}$	40
41	42	43	44	45	$5\bar{4}$	$5\bar{3}$	$5\bar{2}$	$5\bar{1}$	50
51	52	53	54	55	$1\bar{4}\bar{4}$	$1\bar{4}\bar{3}$	$1\bar{4}\bar{2}$	$1\bar{4}\bar{1}$	$1\bar{4}0$
$1\bar{4}1$	$1\bar{4}2$	$1\bar{4}3$	$1\bar{4}4$	$1\bar{4}5$	$1\bar{3}\bar{4}$	$1\bar{3}\bar{3}$	$1\bar{3}\bar{2}$	$1\bar{3}\bar{1}$	$1\bar{3}0$
$1\bar{3}1$	$1\bar{3}2$	$1\bar{3}3$	$1\bar{3}4$	$1\bar{3}5$	$1\bar{2}\bar{4}$	$1\bar{2}\bar{3}$	$1\bar{2}\bar{2}$	$1\bar{2}\bar{1}$	$1\bar{2}0$
$1\bar{2}1$	$1\bar{2}2$	$1\bar{2}3$	$1\bar{2}4$	$1\bar{2}5$	$1\bar{1}\bar{4}$	$1\bar{1}\bar{3}$	$1\bar{1}\bar{2}$	$1\bar{1}\bar{1}$	$1\bar{1}0$
$1\bar{1}1$	$1\bar{1}2$	$1\bar{1}3$	$1\bar{1}4$	$1\bar{1}5$	$10\bar{4}$	$10\bar{3}$	$10\bar{2}$	$10\bar{1}$	100

6. We can note that the above counting table avails the following numerals

Zero 0

positive numerals 1, 2, 3, 4, 5

negative numerals $\bar{1}, \bar{2}, \bar{3}, \bar{4}$

7. There is table (1) corresponding to table (1); the table 1 is counting table, while table (1) is reverse counting. It also may be designated as reverse counting table or simply as negative counting.

7. Likewise the tables of ($\overline{2}$, $\overline{3}$ & $\overline{4}$) may be designated as negative tables corresponding to tables (2, 3, 4) respectively.
8. Therefore the essential pre-requisite knowledge for writing any table up till any number of steps is the values of tables 1, 2, 3, 4 & 5 only. The negative tables would follow of their own as negative values.
9. The tables 1 to 100 can be written with the help of the knowledge of tables 1, 2, 3, 4 & 5 as every number from 1 to 100, as is evident from the above counting table no-2, can be written as 1, 2 or 3 digits form, and each digit to flow out corresponding value table.
10. This way, the one digits number namely 1, 2, 3, 4, 5 shall be flowing out tables 1, 2, 3, 4, 5 and the same can be straight a way be written with the help of the previous knowledge.
11. The table for number 14, being a double digit expression, can be written for values in two parts, first as values of table 1 and second as values of table 4, that is negative values of table 4.

12. The flow chart for this table would follow as under:

Step		First Part	Second Part	Value of Step
Step-1:	1 × 6	1	$\overline{4}$	10-4=6
Step-2	2 × 6	2	$\overline{8}$	20-8=12

Step-3	3 × 6	3	$\overline{1}\overline{2}$	30-12=18
Step-4	4 × 6	4	$\overline{1}\overline{6}$	40-16=24

13. The above steps values are worked out, firstly computing the values for the first part as table-1 and for the second part as table 4 as negative values of table 4, and then the simplification. Here, for mental operations, the knowledge of table 4 is presumed.

14. Likewise the table 1$\overline{3}$ can be written as under

Step		First Part	Second Part	Value of Step
Step-1:	1 × 7	1	$\overline{3}$	10-3=7
Step-2	2 × 7	2	$\overline{6}$	20-6=14
Step-3	3 × 7	3	$\overline{9}$	30-9=21
Step-4	4 × 7	4	$\overline{1}\overline{2}$	40-12=28

15. Likewise the tables 1$\overline{2}$ & 1$\overline{1}$ can be written as under:

Table 1$\overline{2}$

Step		First Part	Second Part	Value of Step
Step-1:	1 × 8	1	$\overline{2}$	10-2=8

Step-2	2 × 8	2	$\bar{4}$	20-4=16
Step-3	3 × 8	3	$\bar{6}$	30-6=24
Step-4	4 × 8	4	$\bar{8}$	40-8=32

Table $1\bar{1}$

Step		First Part	Second Part	Value of Step
Step-1:	1 × 9	1	$\bar{1}$	10-1=9
Step-2	2 × 9	2	$\bar{2}$	20-2=18
Step-3	3 × 9	3	$\bar{3}$	30-3=27
Step-4	4 × 9	4	$1\bar{4}$	40-4=36

16. Now if we have to write tables for 96, 97, 98 and 99, then first of all we shall be rewriting them as $10\bar{4}$, $10\bar{3}$, $10\bar{2}$ and $10\bar{1}$.

Then, as these are 3 digits expressions, so each digit shall be making its independent contribution at each step of the table. Accordingly the values of table are to received in three parts.

Illustratively $10\bar{4}$ shall be flowing out, first part as table 1 values, second part as table 0 values and third part as negative of table 4 values. Therefore our knowledge of tables 0, 1 and 4 shall be helping us to flow out the table of 96 as under

Table 96= Table 10$\bar{4}$

Step	First Part	Second Part	Third Part	Value of Step
1 × 96	1	0	$\bar{4}$	100-0-4=96
2 × 96	2	0	$\bar{8}$	200-0-8=192
3 × 96	3	0	$\bar{1}\bar{2}$	300-0-12=288
4 × 96	4	0	$\bar{1}\bar{6}$	400-0-16=384

Table 97= Table 10$\bar{3}$

Step	First Part	Second Part	Third Part	Value of Step
1 × 97	1	0	$\bar{3}$	100-0-3=97
2 × 97	2	0	$\bar{6}$	200-0-6=194
3 × 97	3	0	$\bar{9}$	300-0-9=291
4 × 97	4	0	$\bar{1}\bar{2}$	400-0-12=388

Table 98= Table 10$\bar{2}$

Step	First Part	Second Part	Third Part	Value of Step
1 × 98	1	0	$\bar{2}$	100-0-2=98
2 × 98	2	0	$\bar{4}$	200-0-4=196

3 × 98	3	0	$\overline{6}$	300-0-6=294
4 × 98	4	0	$\overline{8}$	400-0-8=392

Table 99= Table 10$\overline{1}$

Step	First Part	Second Part	Third Part	Value of Step
1 × 99	1	$\overline{0}$	1	100-0-1=99
2 × 99	2	$\overline{0}$	2	200-0-2=198
3 × 99	3	$\overline{0}$	3	300-0-3=297
4 × 99	4	$\overline{0}$	4	400-0-4=396

■

SECTION - III

FOUR BASIC ARITHMETIC OPERATIONS

I

INTRODUCTION

1. Vedic systems organised artifices of numbers on various place value format. Of these, one of various applied values is ten place value format.
2. Here, unless the context is otherwise, the numbers expressions format taken being availed is ten place value format.
3. Being ten place value systems it makes available nine positive numerals as counts one to nine. In addition, replace value format, provides '0' numeral.
4. As such, the basics of the arithmetic operations need be approached first in reference to these ten numerals namely '0,1,2,3,4,5,6,7,8,9'.
5. This in a way to have arithmetic operations of these ten numerals with themselves.
6. For basic arithmetic operations are 'addition, minus, multiplication and division'.
7. These operations have been provided special notational symbols, viz., + for addition, - for minus, x for multiplication, and ÷ for division.

8. The symbol for addition operation (+) is a symbol of a pair of axes, symbolsing and indicating that the geometric format for the operation is as would be manifesting attainment from a given single axes leading to the pair of axes.
9. The symbol for minus operation (-), is a symbol of a single axis, which in reference to the about symbol (+) shall be a reversal from a pair of axes to a single axis, as a reversal of the addition operation, and hence designating as minus.
10. The symbol for multiplication (X) as comparison to the addition (+) is to attain domain placement, for axes, in place of their boundary placement (of plane/surface). This is a shift from boundary to domain; linear order to spatial order.
11. Multiplication as repeated addition, and the symbolic notations in terms of axes indicating a shift from a boundary to domain deserve to be chase to appreciate the unison of the processes as well as their distinguishing features because of which these (addition and multiplication) emerge to be distinct arithmetic operations, as well, as that of arithmetic operations are ultimately of same origin of artifices of numbers accepting counts as their values manifesting as counting with the rule of 'one more than before' and thereby the whole range of values getting sequenced as a single range.
12. Multiplication symbol is segregation around the axis as a pair of parts, with a point to point correspondence along the pair of parts of the second axis. This operation, as such is a reversal

of the form and format which makes it in reference to multiplication, as is minus related to addition.

Addition Of Numerals

1. Vedic system approached numbers as artifices which presumes existence of parallel dimensional format;

   '**one**' as 'single artifice' and there existing 'one dimensional format/one axis/one space',

   '**two**' as 'a pair of artifices' and there existing 'two dimensional format/two axes/two space',

   and so on,

   '**nine**' as 'nine artifices' and there existing 'nine dimensional format/nine axes/nine space'.

   And

   '**Zero**' as of cipher artifice and their existing cipher dimensional format/zero space.

2. The ordering rule of Ganita sutra-1 'Ekadhiken Purevena' / 'one more than before', manifests organisation format for 'addition operation' unfolding sequentially, and it at its first step of attainment ahead 'one' as 'two' as 'two' being one more than 'one' as the same being the value before the operation.

3. This rule with its repeated application manifests in terms of this working rule as sequential addition, a full range of 'counting', sequentially unfolding format with outline expressible as:

3. This format is of continuously increasing range. Ganita Sutra-2 attains transition to a new format of ten place value system it is of the organisation arrangement of 'all from nine and last from ten',

which is to segregate and classify nine positive numerals from the 'zero' being 'place value numeral'. This transition and attainment from Sutra-1 to Sutra-2, is parallel to the transition and attainment from artifice of 1 to artifice of 2/ one space to two space/one space as domain to one space as boundary/circumference which provides organisation arrangement of 9 points coverage as nine counts/nine numerals and the tenth count is to avail cyclic order (*chakravrat*) and the infinite linear range of Sutra-1 format gets transformed into infinite cyclic order of finite counts parallel to the number of numerals supplied by place value system at the base of the cyclic order which here in the context is ten place value system.

4. This cyclic format of nine points coverage of circumference of a circle deserves to be chased for comprehension and appreciation of the way addition operation manifests as a working rule. Stepwise this chase may be on the following lines:

   (1) The positive numerals 1 to 9 shall be taken as the manifestation range supplied by the organisation format of sutra-1.

   (2) The place value numeral '0' is supplied format by the organisation format of sutra-2, as of spatial feature, as much as that it is of beginning placement prior to the placement of first numeral namely 'one' as well as to be of last placement following the numeral 'nine'; at a place of value equal to 'the place value format', which in the present context

is to be 'ten', as a double digit expression '10' with the first numeral as double digit expression to be '01'. With '10' and '01' as a reflection pair, it makes cyclic order a spatial order absorbing opposite orientations and attaining double digit expression for the whole range of numerals as '00', '01', '02', '03', '04', '05', '06', '07', '08', '09'; and the next number in the range as '10'.

(3) This double digit expression, with zero adopting the symbol of hollow circle makes it as that with every step of coverage of numerals range, the progression ahead are to avail second round, as if along the second circle which is supplied in the double digit expressions for the numerals as '00', '01', '02', '03', '04', '05', '06', '07', '08', '09'. In a sequence , three digit expressions, four digit expressions and so on are to be the availability of another round of circular paths as is evident from the expression format for the numerals as of three digits, four digits and so on of higher formats:

'000', '001', '002', '003', '004', '005', '006', '007', '008', '009' '0000', '0001', '0002', '0003', '0004', '0005', '0006', '0007', '0008', '0009'

---

(4) Availing two digit expression format for the numerals, the addition operations for numerals with numerals shall be sequentially manifesting as:

(1) 09 +01=10.

ADDITION

1. **ADDITION** is one of the basic mathematical operation.
2. It is an **ARITHMETIC OPERATION.**
3. Essentially it is the operation of **COUNTING NUMBERS**.
4. The source Sutra governing the operation and supplying the working rule is Ganita Sutra-1 'Ekadhikena Purvena' / 'By one more than the previous one'.
5. The source Sutra governs the operation of addition of going by one more than before at each step and sequentially attaining an ordered set of counting numbers with internal structure as 0+1, 1+1, 2+1,— —.
6. The word formulation '**ADDITION**' permits its chase as '**ADD IT ION**'.
7. '**ADDITION**' as '**ADD IT ION**' means an operation of adding step by step '**ION**' by '**ION**' / 'unit by unit' / 'one by one'.
8. NVF (ADD) = 1+4+4=9, NVF (IT) =9+20=29, NVF

   NVF (ADD IT)=38=NVF (ION).
9. The rule 'By one more than before' for the counting numbers as artifices may better be chased by availing the presumption of artifices about existence of parallel dimensional formats.
10. The counting number '1' presumes the existence of '1-space' / 'line' / 'axis'.
11. The rule 'One more than the previous one' shall be taking us to a pair of axes.

12. This takes us to the designated symbol for 'plus' / '+'; a pair of axes, joined together.
13. NVF (PLUS)=16+12+21+19=68=NVF (JOINT) gives us the insight about this designated symbol '+' as that the addition is attained as a pair of axes are joined together. Here the focus is firstly upon the availability of a pair of axes and secondly upon their joint.
14. This, as such, further gives us an insight as to the artifices of numbers and the dimensional frames of spaces running parallel, as much as that a jump / shift / transition / transformation form 1-space to 2-space is parallel to a jump / shift / transition / transformation from '1' to '2'. And that this jump / shift / transition / transformation is attained in terms of the working rule of the ordering principle of Ganita Sutra-1 'Ekadhikena Purvena' / 'By one more than the previous one'.
15. '2' is a pair '1,1', but it is a single cardinality, like '1' itself is a single cardinality, and this is inherently inbuilt in it with its attainment as 'plus' / 'joint' for the pair of axes getting joined together as a single frame for the special ordered 2-space.
16. This silently shifts from linear order of 1-space to spatial order of 2-space; as powers of 'joint', joining pair of axes as power-2 / 1×1 / 1^2.
17. This shift inherently takes us from 'addition to multiplication'.
18. Along, linear order format of a line / 1-space, the multiplication as repeated addition just manifests elongation /extension of lengths of intervals.

19. However, along spatial order format of a plane / 2-space, the multiplication is repeated addition of powers manifesting additions of dimensions and attainments of higher spaces.
20. NVF (ADDITION)=76=NVF (MINUS), makes these operations of opposite orientations like reaching from artifice 6 to artifice 7 and the reverse from artifice 7 to artifice 6. It would be a blissful exercise to reach from designated symbol '+' for addition as plus to the designated symbol ' – ' for minus.

ADDITION OF TWO DIGITS NUMBER

Sources:

1. Ganita Sutra 1 एकाधिकेन पूर्वेण Ekadhiken Purvena Rule: One more than before.
2. Ganita Sutra 14 एकान्यनेन पूर्वेण Ekanunena Purvena Rule: One less than before
3. Lilawati: अथ संकलित–व्यवकलितयोःकरणसूत्रं वष्ताधर्म्। Here is addition-subtraction method in half verseकार्यः क्रमादुत्क्रमतोष्थवाष्ङ्.कयोगो यथा स्थानकमंतरं वा।।

   Skill steps be done in sequence the operations of addition and subtraction as per the places, right to left, of digits (of numbers to be added/ subtracted)

Case 1: Add 00 and 01

Step 1: Let us write the given numbers pair in a format as:

First number	0	0
Second Number	0	1

Step 2: To add the digits of first slots of both numbers as:

By definition, under rule of Sutra-1

0+1=1 as Ekadhika of 0 that is 1

Step 3: To add the digits of second slots of both number as:

Step 4: Result of addition operation:

First number	0	0
Second Number	0	1
Result of addition operation	0	1

Step 2: To add the digits of first slots of both numbers as:

By definition, under rule of Sutra-1

1+1=2 as Ekadhika of 1 that is 1

Step 3: To add the digits of second slots of both number as:

By definition, under rule of Sutra-1

0+0=0 for want of Ekadhika

Step 4: Result of addition operation:

First number	0	1
Second Number	0	1
Result of addition operation	0	2

Cases for exercise:

1. Write down the result of addition operation for the following case:

First number	0	1
Second Number	0	1
Result of addition operation	0	2

2. Write down the result of addition operation for the following case:

First number	0	1
Second Number	0	3
Result of addition operation		

Case 3 add 02 and 02:

Step 1: Let us write the given numbers pair in a format as:

First number	0	2
Second Number	0	2

Step 2: To add the digits of first slots of both numbers as:

By definition, under rule of Sutra-1

2+2=4 as Ekadhika of Ekadhika of 2 is 4.

Note: Here the Ekadhika rule is applied in two steps, firstly it takes from 2 to Ekadhika 2 that is to 3 and then in the second it takes to Ekadhika of 3 that is to 4

Note: Here the numeral 2 has been chased for its addition with two on first principles of Ekadhika rule. It would be good exercise to chase addition of digits on first principles.

Step 3: To add the digits of second slots of both number as:

By definition, under rule of Sutra-1

0+0=0 for want of Ekadhika

Step 4: Result of addition operation:

First number	0	2
Second Number	0	2
Result of addition operation	0	4

Cases for exercise:

1. Write down the result of addition operation for the following case:

First number	0	2
Second Number	0	3
Result of addition operation		

2. Write down the result of addition operation for the following case:9th place

First number	0	3
Second Number	0	3
Result of addition operation		

3. Write down the result of addition operation for the following case:

First number	0	4
Second Number	0	5
Result of addition operation		

Case 4 add 05 and 05:

Step 1: Let us write the given numbers pair in a format as:

First number	0	5
Second Number	0	5

Step 2: To add the digits of first slots of both numbers ..as:

By definition, under rule of Sutra-1 5+5= as Ekadhika of Ekadhika of Ekadhika of Ekadhika of Ekadhika is 10. Here the Ekadhika rule is applied in five steps, firstly it takes from 5 to Ekadhika 5 that is to 6 and then in the second step to 7, in third to 8, in fourth step to 9 and finally in fifth step to 10

Further as 10 is Dashadhika of zero:(.0), as such in two digit expression it accepts zero as digit in first slot and one as a digit in the second slot.

Note: It is this situation where sum of digits become Dashadhika numeral, the expression for it as two digits takes digit "1" in the second slot. With it the second slot gets one additional count. It is known as carried forward count. Because of this carried forward count, the sum of the digits of second slot, under the rule of Ekadhikena, makes the value of the sum of digits of second slots as of one count more.

Note: The sum of two numerals, under the above process can yield only "1" as carry forward count for the next slot.

Step 3: To add the digits of second slots of both number as:

By definition, under rule of Sutra-1
0+0=0 for want of Ekadhika

Step 4: Result of addition operation:

First number	0	5
Second Number	0	5
Step-1 10 as Dashadhika of 0	0	.0
Step-2 expression for Dashadhika	0	10
Step-3 carry forward for one	0+1	0
Step-4 Ekadhikena of 0	$\dot{0}$	0
Result of addition operation:	1	0

Cases for exercise:

Add 05 and 06:

Hint:

06 = 1 $\bar{4}$

First number	0	5
Second Number	1	$\bar{4}$
	1	1

Add 05 and 07:

Hint:

06 = $1\bar{3}$

First number	0	5
Second Number	1	$\bar{3}$
	1	2

Add 05 and 08, 05 and 09:

Case 5 Add 37 with 73:

Hint: First slot digits add 10 and give "one as a carry forward count for second slot". The second slot digits also add 10 but carry forward count from first slot will make the sum as Ekadhikena that is 11. This two digit expression would suggest that from here their should one carry forward count for the next slot. As a result, the sum of 37 and 73 would require a format of three slots so that one as carry forward from second slot can be given placement. The general skill of addition for numbers of any number of digits shall be taken up during following aspects. However from the above, one may have an idea about the nature of the skill of addition operation for numbers.

SUBTRACTION

1. '**SUBTRACTION**' is another basic arithmetic operation.
2. It is of opposite orientation than that of '**ADDITION**'.
3. The **source sutra** governing this operation and

supplying the working rule is Ganita Sutra-14 'Ekanunena Purvena' / 'One less than before'.

4. 'Addition' is 'add it ion', while the subtraction is ' subtract ion'.
5. NVF (ADDITION) = 76= NVF (MINUS).
6. The designated symbol '+' for addition with subtraction operation, of 'subtract ion' takes to the designated symbol '-' for it as the pair of axes (+) taking us to a single axis (-).
7. Ganita Sutras take up the phase and stage of simultaneous happenings of 'addition and subtraction' as Ganita Sutra-7 'Sankalana-Vyakalana-bhayam'.
8. Illustrative situation of this happening of geometric formats is that of a line simultaneously accepting a pair of orientations.
9. The lines with orientation features, may be designated as of positive and negative orientations respectively, and expressed as positive and negative artifices, symbolized as (+1) and (-1) parallel to which would be (+1 space) and (-1 space) respectively.
10. The (+1 space) and (-1 space) as per Vedic systems comprehension and approach to manifested reality of linear order as manifestation layer accepts and approaches (+1 space) with (-1 space) playing the role of its dimension.
11. As such, the simultaneous happening of 'addition and subtraction' is the basic happening whose comprehension and chase is of basic focus.
12. This simultaneous happening of 'addition and

subtraction' is of big range of applied values. The word formulation 'Sankalana' means accumulation / compilation /earning / addition and 'Vyakalana' means spending / subtraction and the like. As such the whole range of happenings of opposite orientations like 'pulls and pushes' from within the 'orb of the sun', and the activities of the commercial world, statistical domain, probabilities range, the life phenomenon of simultaneous increase and decrease of age, and the like is the coverage range of this phase and stage of simultaneous happening of 'Sankalana-Vyakalana-bhayam'.

13. Ganita Sutra-7 may be approached for its different format feature as :

13.1.1 Simple English rendering are the working rule of Ganita Sutra -7 is to process "by addition and subtraction".

13.1.2 The processing here is in terms of a pair of processes, firstly by addition and secondly by subtraction.

13.1.3 Addition is an independent process; infact the first arithmetic operation of classical systems of mathematics.

13.1.4 Likewise subtraction is another independent process.

13.1.5 As such the form, Ganita Sutra-7 for its organization format and further for its working process, is taken up, here at introductory stage, the pair of processes availed here namely addition and subtraction, for their independent existence as well as for their simultaneous existence are being introduced.

MULTIPLICATION

The timely care for smooth transition to geometric formats may require to draw horizontal line and take it first (line or axis) and then to draw second line (as second axis). This first and second [line (s)/axis (axes)] will help us shift the mathematics (of counting and also of arithmetic and other operations from line to a plane, which is going to be really a very big break through which is ultimately to help us remove many of the bottle necks as well as the mental blocks because of linear mathematics/mathematics of line format).

The horizontal line progression is to be availed as a format for addition under the rule of "one more than before". The second/ "Vertical line" is to be availed (presuming the existence of the first line/ "Horizontal line") for the format for multiplication (under the rule of one more than before and the rule of vertically and crosswise).

STEPS FOR MULTIPLICATION SKILL

Skill now being introduced is of multiplication. For it we shall begin with construction of the format as:

(i) Write 1 as 0+1 as entries of the digits along two slots arranged horizontally as:

0	1

(ii) If we have to multiply 1 with 1 then the first 1=0+1 is written as first row and the second 1=0+1 is written as second row.

0 1

0 1

(iii) For multiplication result, another row of three horizontal slots is constructed and in terms of this the format formation emerges to be as:

0 1

0 1

(iv) The three steps in which the three parts of the multiplication are reached at are:

First step: To multiply the first part of upper row that is 0 with first part of the lower row that is zero.

$$0 \times 0 = 0$$

Second step: Multiply first part of upper row that is 0 with second part of the lower row that is 1 and add the product of second part of first row with the first part of the lower row.

$$0 \times 1 + 1 \times 0 = 0 + 0 = 0$$

Third Step: To multiply the second part of the upper row that is 1 with the second part of the lower row that is 1.

$$1 \times 1 = 1$$

(v) Add the first second and third steps values.

$$0 + 0 + 1 = 1$$

(vi) This way the expression on the format would be

Learning and teaching of the skill of multiplication up till 5x5.

(0×0, 0×1, 1×1, 0×2, 1×2, 2×2, 0×3, 1×3, 2×3, 3×3, 0×4, 1×4, 2×4, 3×4, 4×4, 0×5, 1×5, 2×5, 3×5, 4×5, 5×5).

Vedic systems presume the learning and remembering of multiplication tables up till 5x5 for further mental computations. As such the students to learn with deep love for kindness of the teacher taking every care to perfect this skill of working out the multiplication tables up till 5x5 as the basic talent requirement for mental mathematics and other mathematical activities, skills, proficiencies and perfection of intelligence.

The exposure and initiation for grasping of the working steps and firm comprehension for perfection of this skill is to be in terms of the above format expression as the first step of learning.

Then as a second step the above illustration be chased step wise. Thereafter all other tables are to be taken up one by one.

Here the solemn duty of the parents is to ensure that the child has in fact worked out all the above multiplication tables with each step correctly.

Here are drawn the outlines of the format for 1x2 and 2x3. Details may be filled:

Outline of format for 1x2

0	1
1	1

0×1	0×1 + 1×1	1×1

Note: We know 0x1 as well as 1x1 and also 0+1.

Outline of format for 2x3

1	1
1	2

1×1	1×2 + 1×1	1×2

Note: We know from above 1x1, 1x2 and also 3=2+1

Exercises:

Multiply 1×3, 3×3, and multiply numeral x numeral for all values of numerals.

5.3.6 MULTIPLICATION OF TWO DIGITS NUMBERS

Linking Note:

A word with teachers

The aspect 01 had availed the source of Ganita Sutra-1 (एकाधिकेन् पूर्वेण) Ekadhikena Purvena (One more than before) as concept for the format of counting. The conceptual skill introduced for reaching at tables up till 5x5 had further availed the source of Ganita Sutra -3 (ऊर्ध्वर्तियकभ्याम्) Udharavtriyakvyam. That the final format reached at for learning the skill of tables up till 5x5, as arithmetic operation of multiplication, accepts the following format:

A	B
C	D

ACCEPTANCE	AD+BC	BD

This format, for multiplication, has been available, and in fact has been availed for multiplication of numerals, with start with counting format of a line with addition as a rule, for progression of counting along the line.

Now this very format, with proper adaptation, is being again availed here for the present conceptual skill aspect 02 for multiplication for the numbers expressed in ten place value system. In fact, this format can be availed for multiplication for the numbers expressed in any place value system but here this format is being adapted in the present aspect of conceptual skill of multiplication of numbers expressed in ten place value system.

First, naturally, it is to be ensured that the students being exposed for learning of the present conceptual skill, are knowing the basic features of numbers expressed in ten place value system in terms of nine numerals (1,2,3,4,5,6,7,8,9) and tenth special digits that is 0, available because of 10 responsible for making the system for the expression of numbers as ten place value system.

Now it is being proceeded with this aspect, presuming that the students have already learnt the special features of expression of numbers in terms of numerals 1,2,3,4,5,6,7,8,9 and special digit 0. Here the most basic feature is that the values within the expression for the numbers depends not only upon the values of numerals and special digit but also upon the placements. Illustratively the number twelve when is expressed as 12, here for the last place numeral 2, the place value being 1, as such its number value at this place comes to be 1x2=2. However for the first place numeral 1, the place value being 10, so its number value at this place comes to be 1x10=10. The total number value for number 12 comes to be 10x1+1x2.It is because of difference in place values (which for ten place value is in the order of powers of ten which as per convention of writing, increases from right to left; and the digit at the extreme left is by convention taken as the last digit at a place with value of 0 power of 10 that is 1.

In this background, we proceed for adoption of the

above format for multiplication for two digits numbers (expressed in ten place value system).

Adoption steps of the above format:

(i) Write 11 as 10+1 for entries of the digits along two slots arranged horizontally as:

1 1

(ii) If we have to multiply 11 with 11 then the first number 11=10+1 written as first row and the second number 11=10+1 written as second row and both rows written one blow another, would express as:

1 1

1 1

(iii) For multiplication result, another row of three horizontal slots is constructed and in terms of this the format formation emerges to be as:

1 1

1 1

(iv) The three steps in which the three parts of the multiplication are reached at are:

First step: To multiply the first part of upper row that is 1 with first part of the lower row that is

$$1\times1=1$$

Second step: Multiply first part of upper row that is 1 with second part of the lower row that is 1 and add the product of second part of first row with the first part of the lower row.

$$1\times1+1\times1=1+1=2$$

Third Step: To multiply the second part of the upper row that is 1 with the second part of the lower row that is 1.

$$1\times1=1$$

(v) Here the format adaptation makes the expression for the result of multiplication of 11x11 as a number expressed in ten place value system by taking three parts in given three slots as values at those places in ten place value system for the number as 121.

Note:

1. The number 121 is of value $1\times10^2+2\times10^1+1\times10^0$.
2. For the present it be taken that $10^0=1$.

(vi) This way the expression for the format after adaptation for multiplication of two digits numbers of ten place value system, in the context of the illustrated case of 11×11 be taken as:

1	1
1	1

1×1	1×1+1×1	1×1
1	2	1

Learning and Teaching of the Skill of Multiplication of Two Digits Numbers.

First group

(i) 10×10, 10×11, 10x12, 10x13, 10x14, 10x15, 10x16, 10x17, 10x18, 10x19

(ii) 11×11, 11×12, 11×13, 11×14, 11×15, 11×16, 11×17, 11×18

(iii) 12×12, 12×13, 12×14

(iv) 13×13,

Note:

Here in the above choice, caution has been that, values for all the three slots remain up till 9 so that the need for carrying forward the value from first or second slot to second or third slot does not arise and within third slot also the values remain a single digit value. This precaution here is simply to avails illustrations, which can be worked without necessity for carrying forward. And that the situations of carrying forward to be taken up with specific focus after learning the skill on its steps as such of the adopted format. The transition from aspect 01 of previous day learning to present aspect 02 deserves to be consciously cognized and learnt, rather to be comprehended well.

Note:

Here are drawn the outlines of the format for 11x12 and 12x13.

Details may be filled:

Outline of the format for 11×12

1	1
1	2

1x1	1x2+1x1	1x2
1	3	2

Outline for the format for 12×13

1	2
1	3

1x1	1x3+2x1	2x3
1	5	6

The above multiplication exercises for cases of first group be undertaken for ensuring oneself about the steps of this skill.

DIVISION

INTRODUCTORY

Division is one of the basic arithmetic operations. Its skills are based upon the basic features of whole numbers. Division is also viewed as reverse of multiplication. Multiplication can be viewed as repeated addition. The addition and subtraction are viewed as reverse of each other. Like that all the four basic operation ultimately get connected as an interconnected operations availing features of the whole numbers.

Nikhilam Sutra has been availed for Multiplication as well as for the Division. The Nikhilam Sutra निखिलं नवतश्चरमं दशतः (Nikhilam Navatascaramam Dasatah) literally translated, (Sutra) means: "all from 9 and the last from 10"

The '**Division**' has been approached in many different ways by the Ganita Sutras. One of the approach to the division operation is in term of Nikhilam Sutra.

The first concept, which need be comprehended well in the context is to have interconnection of the divisor with the base the First case of the situation is in respect of the numerals (9, 8, 7, 6, 5, 4, 3, 2, 1) as divisors. These numerals are interconnected with base 10 as (10-9=1), (10-8=2), (10-7=3), (10-6=4), (10-5=5), (10-4=6), (10-3=7), (10-2=8), (10-1=9). These relations may be viewed as that 9 is deficient by 1 from the base 10. Likewise 8 is deficient by 2 from base 10 and so on.

Once the numeral as divisor is viewed as to how much deficient it is from the base, then the division of any number expressed in ten place value system can be taken as of two parts, the last digit of the number and the remaining part of the number. The smallest two digit number is 10. Its first part (last digit) is '0'. The remaining part of this number is '1'. This number can be expressed as 1x10+0. When we have to divide it, say by 9, we can rewrite it as [1x(9+1)+0] we can see that on division by nine the second part of the number, that is, 1x(9+1)=1×9+1x1=9+1 shall be on its division by 9 be yielding 9/9 + 1/9=1+1/9. Here 1 is quotient. This 1/9 is carried forward to unit place as 1 which well added to first part of the number, namely, '0', it sums up 0+1=1 as remainder.

Likewise if we have to divide 11 by 9, the same can be worked as second part being 10 so it shall be yielding '1' to be carried to the first part to workout remainder as 1+1=2. Hence quotient would be 1 and the remainder would be 2.

If we have to divide 12 by 9, its second part '10' again shall be yielding 10-9=1 to be carried forward to the first part to workout remainder being 2+1=3.

If we have to divide 13 by 9, its second part '10' again shall be yielding 10-9=1 to be carried forward to the first part to workout remainder being 3+1=4.

If we have to divide 23 by 9, its second part '20' shall be yielding (10-9)+(10-9)=2 to be carried forward to the first part to workout remainder being 3+2=5.

The working chart may be drawn for divisions by 9 as under.

Step-1 Divide the number, say 11, into two parts, while separating the digits of this number by a vertical line to express as

Second part		First part
1	|	1

Step-2 Put divisor (9) on the left hand side by separating the above expression for the dividend (11) by a bracket as

	Second part		First part
9)	1	|	1

Step-3 Working step-1

(i) The digit of the second part above, that is (1), is to be noted down as 'quotient', and further.

(ii) The digit of the second part above is to be carried forward to the first part for computation of remainder,

And, the working chart would progress accordingly as under

	Second part		First part
9)	1	|	1
	______	|	______1_ (brought from second part)

■

SECTION - IV

GEOMETRY CLASS IN AYUR VEDIC GARDEN

1
INITIATION OF YOUNG MINDS

1. Vedic teachers initiate young minds about Vedic geometry by taking their class in Ayur Veda garden for the reason that Ayur Veda garden is the complete package of required environment because of two prominent features of Ayur veda garden as that :

   (i) It has Sathapatya format, and

   (ii) Plants collectively are complete Ayur veda package availing Sathapatya format.

2. The Sathapatya format of Ayur veda garden is there to, well receive, the jyoti transcendence from rays of Sun. And, the Ayur Veda package plantation is there to attain full Sama ascendance to make the Space / Akash 'transcendental' and to reach the orb of the Sun through rays of the Sun.

3. Teacher accepts this set up for the geometry class being most appropriate for initiating young minds for Vedic systems in general and Vedic geometric system in particular because of the complete unison between the Jyoti transcendence through

rays of the Sun and Sama ascendance through rays of the Sun.

4. Jyoti, colours; Sama, frequencies; Sathapatya, formats; frames and forms are all there for initiation of young minds for Vedic geometric systems about Reality; existence within frames. And, naturally, teacher avails all these for proper initiation of young minds. Without making conscious, and before formally declaring formal initiation of education of Vedic geometric systems, teacher takes the students along with him for a round of the garden and stops at every Sathapatya format zone, with reaching at, a new Ayur vedic plant. Teacher takes the students along to every storey of the garden and familiarizes the students with everything there in the garden keeping in mind that he is to avail all these impressions at the time of formal education instructions of the Vedic geometric systems.
5. Curiosities about existence within human frame, Pursha format, food to medicine requirement, space to source urge of students to know about all for their complete answers is also there inherent in the curiosities raising pre formal instructions round of the Ayur Veda garden under taken by the teacher along with students.
6. Sathapatya, Sankhya, Ayur Veda; Yajna, Yoga, Yag; Jyoti, Sama, frequencies; transcendental, self referral and unity state consciousness impulses are all there which come into play during this round of the Ayur Veda garden and it is this ambrosia of bliss all round about vibrating life in

Ayur Veda Plants all round which is going to be the attainment needed by the teacher for its formal initiation instructions for Vedic geometric systems.

7. The ambrosia of bliss all round, the young minds with this round of Ayur Veda garden is all what *Ativahkas* (Brahman intelligence embedded carriers within rays of the Sun) shall be waiting for giving ride to young minds for intellectual pilgrimage to the transcendental worlds awaiting ahead for their welcome in the Vedic geometry class.
8. The new wonderful worlds of different orders and new wonderful technologies accepting linear, spatial, solid and hyper solid orders are awaiting young minds coming to geometry class in Ayur Veda garden and young minds with round of the Ayur Veda garden shall be vibrating, full of ambrosia to satisfy their urge about new wonderful worlds of Vedic geometric systems.

■

To know more

Ved means knowledge. It also means to know and to reach at knowledge. Originally there was 'one Ved'. Later on this knowledge was re-orgnaised as four Vedas, as four sets of pure knowledge. Corresponding to teach set of pure knowledge has been set of applied values, designated as UpVedas, Four Vedas are Rig Ved, Yajur Ved, Sam Ved and Atharav Ved. Four Up Vedas are Ayur Ved, Dhanur Ved, Ghandarav Ved and Sathapatya Ved.

■

2

MANSARA (ESSENCE OF MEASURE)

1. Each Ayur Vedic plant has its unique mathematics, science and technology of receiving and availing Panch Mahabhut and Jyoti from rays of the Sun and simultaneously manufacturing and flowing out Sama for the consumption of Panch Mahabhut and orb of the Sun.
2. As such, each Ayur Vedic plant has its own specific packages of requirements, first of which is the Sathapatya format for its plantation. This in fact is format, of surface, may be at ground, are any of the upper storeys; or of pots, may be for ground placements or for their hanging positions. This initial requirement's approach has been addressed by Sathapatya, and same is designating as answerable in terms of measures. This that way becomes the essential feature of applied values of Sathapatya and as a result there is *'MANSARA'* (Scriptures, of Sathpatya order); *Man* means measure and *Sara* means essence, and as such it is 'essence of measures' which is essentially being addressed here in the said scripture of Sathapatya.

3. Accordingly, the teacher of geometry has a scriptural command here that the formal instruction of initiation of young minds for discipline of Vedic geometry (based on Sathapatya) is to well keeping in mind that it is 'essence of measure' which is the most fundamental feature around which the initiation for perfection of comprehension of this discipline is to be achieved. With this the focus of instructions for formal initiation of young minds for the discipline of Vedic geometry systems based on Sathapatya order is to be the "Measure".
4. This focus and initiation, with this focus, is in fact to perfect the intelligence of young minds for the whole range of Vedic mathematics which at initial stage is 'Arithmetic to Astronomy' and that which at its next stage is to be from 'Sankhya to Jyotish' and at its final stage is to be of the transcendental phenomena with transcendental world (5-Space) itself to be in the role of dimension, Sun in the role of boundary, Sapat Rishi Lok as the domain, Asth Prakriti as origin and Nav Braham as the transcendental flow source leading to Par-Braham.
5. As such, the first instruction for the young minds in the classic geometry in Ayur Veda garden is to be three fold "*Aaan* (आन)] *Maan* (मान) and *Shaan* (षान). It is measure (*Maan*/मान), for which young minds approach with respect (*Aaan*/आन) and go glorious and transcendental (*Shaan*/षान) after its (measure's) comprehension.
6. It is like this the young mind with respect (*Aaan*/

आन) perfects intelligence with full comprehension of measure (*Maan*/मान) and in the process mind of perfected intelligence goes glorious and transcendental (*Shaan*/शान) for his whole existence.

■

To know more

Sathapatya Ved is Up Ved of Atharav Ved. Mansara is scripture of Sathapatya Ved.

Atharav Ved has nine branches. Every branch of Ved has distinct Samhita, Brahmana, Arnik, Upnishad and Upved.

As per this organisation, their are nine distinct branches of Sathapatya Up Ved.

The present day Disciplines of knowledge designated as mathematics, sciences and technologies, on first principles, are to take us to 'Sathapatya Up Ved' of Atharav Ved.

The first quarter of first mantra of Atharav Ved enlightens us as that 'Yeh Trishapta Priyante Vishwa' / This world is enveloped by Trishabta / paired artifices 'three and seven'.

The amongst others, the pairing of artifices '3 and 7 shall be yielding seven versions of cube . seven geometries of 3-space. It also shall be accepting one of the pairing situaitons being '37'.

NVF (CONE) = 37 which on its pairing with itself shalll be leading to 'NVF (CONE, CONE)= 74= NVF (PAIRING)'

■

3
COUNTS AND UNITS

1. Measure, may initially, better be introduced as counting.
2. Simultaneously measure be introduced through scale units by availing counting for counting units.
3. The counting with the help of plants in a row, branches of a plant, leaves in a branch, fruits along a branch, seeds in a fruit and like are to perfect the intelligence regarding counting.
4. Counting units along a scale, the number of times scale availed for measuring the length of the row of a plant, all the rows of a field and the like are to perfect the intelligence regarding units (marked along a scale) with the help of counting.
5. The counts and units as such be introduced as Mathematical parallel tools.
6. The branches of the trees be got measured with the help of scale and the units be got tabulated to have comparison of the lengths of the branches of a given plant.
7. Likewise the largest length of one plant be got

measured with the largest length of the branch of another plant.

8. This exercise of comparison of length of different branches may be had in different situations which may be to have comparison of lengths of any two branches of a plant or to have comparison of the lengths of any given branch of one plant and any branch of the second plant.
9. The stem, branches, leaves and the like may be the thorns, or any of them may be subject of measurements with scale.
10. The concept of surface may be introduced in the background of the concept of length accepting scale measures.
11. The surfaces of leaves, branches, stems may be availed to introduce the concept of surface and of area units.
12. Like wise the concept of volume may be introduced in the background of concept of length and area, line and surface in the context of the branches, stems, fruits and the like other solids, surfaces and lines marking their presence in abundance in the Ayurveda garden. ■

To know more

NVF (UNIT) = 64 = NVF (ZERO).

This feature of pairing operation of spatial order of creator's space (4-space) deserves to be chased as the linear units of 3-space emerge as zero units at boundary of 4-space. ■

4

LINES AND CIRCLES

1. Teacher shall expose young minds to the formats, frames and forms of different Ayur Vedic plants and different parts of individual plants.
2. Simultaneously the teacher shall focus young minds to the Sathapatya formats of ground zones, field shapes, plantation organization within the Ayur Veda garden.
3. As a practical demonstration and for shaping young minds for practical approach, the rope, scale, pegs may be availed for focus about classification of geometric forms in different categories like points, edges; lines, thorns, surfaces, leaves; solids, fruits and the like.
4. For firm comprehension of different stories lay out of Ayur Veda garden, as a first step, teacher shall help distinguish between line and surface.
5. For that, the teacher shall help students to draw a straight line of given length and to draw a circle of given dia-meter.
6. For ensuring achievements of this skills of drawing

lines of given length and for drawing circle of given dia-meter, the different fields within the garden availing such formats for rows of plants and for the circular year marked zones for the plants, individual or collectively, may be got measure.

7. As an exercise, the student may be asked to draw a line for given length starting from a given point.
8. Further as an exercise the student may be asked to fix the middle point of given line with the help of a thread/rope.
9. The middle point of given line may be availed as a center for the circle of given dia-meter.
10. With given center and given rope of half length of dia-meter, student may be asked to draw a circle and to measure the lengths of two lines with one point for both being the centered and other ends for both lines be on the circumference of the circle.
11. Fields in the Ayur-Veda garden may be spotted which are circular in shape and there is Tulsi plants at the center/origin.
12. The concept of Angle may be introduced.
13. The concept of Arc and chord may be introduced.
14. The concept of Tri-angle within the circle may be introduced.
15. The concept of ratio of lengths of two lines may be introduced.
16. The concept of ratio of lengths of two sides of tri-angle may be introduced.
17. The skill of division of circular disk into four parts may be introduced.
18. The concept of circumference of circle with unit dia-meter may be introduced.

19. The special unit for the circumference of a circle with unit dia-meter may be introduced.
20. The skill of measuring the circumference with the help of thread/rope as whole number of units may be introduced.
21. It be demonstrated that the length of the thread covering circumference of circle of unit length is more than three units.
22. It be demonstrated that the length of the thread covering circumference of circle of ten units length is more than thirty one units.
23. The Up Sutras 1 & 2 may be introduced for lines of proportionate lengths and circumferences of circles of proportionate dia- meters.

■

To know more

The creator's space geometry being at work in the set ups of trees deserves to be chased as that:

NVF (TREE) = 48 = 2 × 4 × 6, and

NVF (TREES) = 67 = NVF (WATER)

And that Water is the second element (the first being Earth). Further that 2 as 1 shall be making the dimensional order for first, first two and all the three dimensions to be collectively as 2x4x6.

The artifice 48 can be chased 8 at first place parallel to 8 solid boundary components being of lienar order and 4 as hyper soild doman of spatial order with second placement. Still further the artifice 67 can be chased as seven edges to six surfaces.

■

5

CUBE

1. The first feature of cube which deserves to be comprehended is that it is a geometric envelop, a nature's bag for carrying matter which is responsible for its volume.
2. This geometric envelop, as it is stitched geometrically is having following components:

   (i) 8 corner points (ii) 12 edges (iii) 6 surfaces

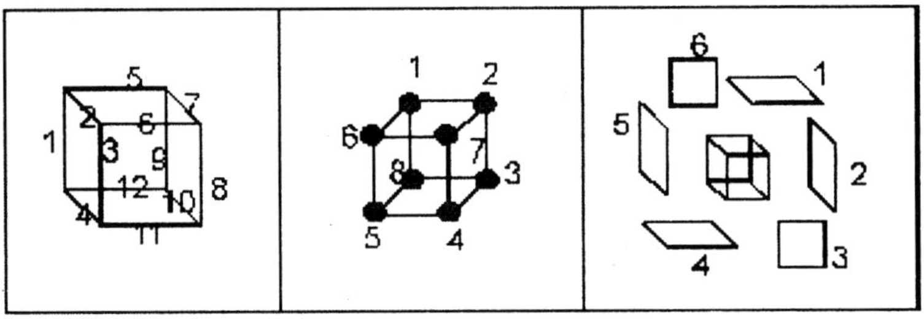

3. As such, as first phase, these twenty six components constitutes its features responsible for providing this geometric envelop, eight corners, twelve edges and six surfaces.
4. The second phase of its features which deserves to be exposed, as that it has:

(i) Length (ii) Breath (iii)Height

but all the three, length, breath and height being equal.

5. It would be good practical exercise to cut out cubes from the wood of braches of Tulsi plant and other Ayur Vedic plants.
6. It also would be very good exercise to make cubes of clay.
7. Cubes of wood, clay and of other materials and of different sizes be prepared.
8. Ice cubes are very interesting cubes.
9. In terms of ice cubes, the concept of solids contents converted into liquid contents may be introduced.
10. This may help have transition from the concept of element of Earth to element of water.
11. The concepts of seeing, smelling, touching, feeling hot and cold, feeling frequencies of vibrations of leaves of different Ayur Vedic plants, as solids, as surfaces and as sources of formats, frames, forms, also can be introduced in terms of the concepts of geometry envelops, bags, containers, carriers of nature and the smell and vibrations emanating there from.
12. The concept of flow of content through roots to stem, branches, leaves, flowers, fruits also can be introduced.

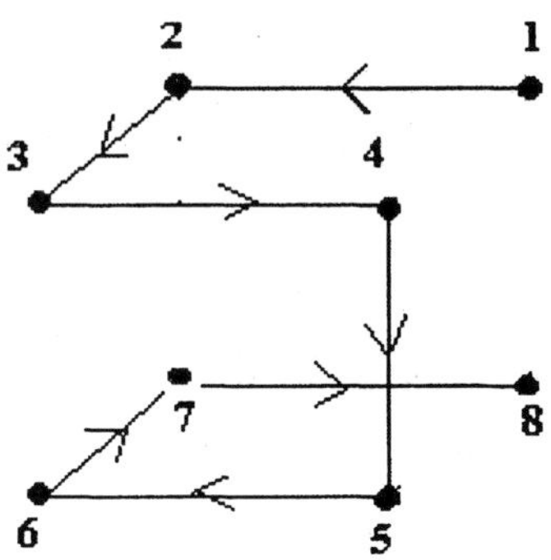

13. The concept of measure as flow of content, duration of flow and the phenomena of motion may be introduced.
14. The concept of static and dynamic positions, the concept of flow, motion; the concept of outer (geometric structure) and inner content structure may be introduced. Cube as geometric envelop of twenty six component and the cube as content as volume are two different geometric set ups as much as that volume is their because of length breath and height while the geometric envelop components are just points, edges and surfaces of length and breadth only.

■

To Know More

A smooth transition deserves to be had from a cube to a cuboid. It is to be in continuity of a transition from a square to that of a rectangle. The transition from a square to that of a rectangle is to be had in continuity of the interval being the specific expression for the whole 3-space being the general expression. In the context, it would be a blissful exercise to chase all intervals being of same specific features as specific expression of 3-space.

A step further shall be taking to a zero space where all points are already blissful in affine state to be the same in all respects. A reverse order chase from equal points to un equal intervals as well being of same specific values and then the squares and rectangles, as well as cubes and cuboid emerging to be complete ranges of specific values of 2 and 3-space respectively.

■

6

BODIES IN MOTION

1. The concept of static and dynamic bodies be introduced.
2. The dynamic bodies may be introduced by throw of balls in the space; branches of tree remaining intact with the stem but still moving in space during storm.
3. Line may be introduced as a track of a moving point. This is one point fixation of a line.
4. Line also may be introduced as length between given two points. This is a pair of points fixation of a line
5. Line further may be defined as every pair of points on it having another point in between the given points also being along the line. This is a fixation of line in terms of triple points.
6. Plane may be defined as moving line. This is a fixation of a plane in terms of single line.
7. Plane may be defined in terms of length and breath. This is a fixation of plane in terms of pair of lines (as length and breath).

8. Plane may be defined as a surface enclosed within three lines (a triangle). This is a fixation of in term of three lines.
9. Cube may be defined in terms of a moving plane (square). This is a fixation of a cube in terms of single plane (Square).
10. Cube may be defined in terms of pair of planes covering specific space (3-Space/Cube). This is a fixation of a cube in terms of pair of planes.
11. 3-Space accepts octants cut in terms of three planes along three axes (parallel to length, breath and height). This is a fixation of cube in terms of three planes.
12. This three ways fixations of interval (line), Square (plane) and Cube (solid) inter connects the geometric set ups of static and dynamic positions of line, surface and solids. The other way, these three geometric set ups of interval square and cube admit inter connection is in terms of the geometric envelops and contents carried as close interval, square and cube may be depicted as tabulated below:

Close interval		Square		Cube	
End points	2	Corner points	4	Corner points	8
Content (length)	1	(Boundaries ines)	4	Edges	12
		Content (Area)	1	Surfaces	6
				Content (Volume)	1
Total	3	Total	9 = 3 × 3	Total	7=3x3x3

■

7

CREATOR'S SPACE

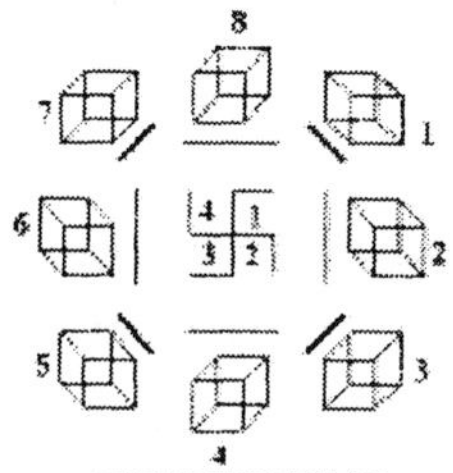

CREATOR'S SPACE

1. Point (as 0-Space body / •) in motion accepts 1-Space/ — as its track.
2. Line (as 1-Space body/ —) in motion accepts 2-Space / □ as its track.
3. Plane (as 2-Space body / □) in motion accepts 3-Space / ▣ as its track.
4. Then as we are seen solids (as 3-Space / ▣ bodies) in motion, their tracks naturally are to be of 4-Space ⌑.
5. 3-Space / ▣ has three dimensions. This is parallel to our Vishwa (3-Space / ▣) having three heads.
6. 4-Space / ⌑ has four dimensions. This is parallel to Lord Brahma, four head lord, creator the supreme.
7. As such 4-Space / ⌑ is creator's space.
8. Each head of Lord Brahma is having a pair of eyes.

8. As such creator's space is a spatial order space.
9. Spatial order, as comparison to linear order of our vishwa, may be defined as an interval of two parts; an interval of two units length, as comparison to linear order as an interval of single unit.
10. The interval of two units may be introduced as a di-monad, a monad of two parts parallel to head with a pair of eyes.
11. Lord Brahma, four head Lord Creator the supreme sits majestically on a lotus seat of eight petals. The Idol of Lord Brahma creator the supreme accepts format of representative regular body of 4-Space (which may be designated as hyper cube-4).
12. It would be an interesting exercise to reach at the format of hyper cube-4 in continuity of the formats of close interval, square and cube.
13. The spatial order (that is plane) as dimension (axes) parallel to head with a pair of eyes makes human frame as of one head with a pair of eyes as a quarter of Brahman, Swastic pada, and with it the applied values needed for bliss full existence within human frame stands settled as such of acquiring enlightenment quarter by quarter of the Brahman domain. The Ayur veda package, as Up Ved of Rig Ved, as one out of four Vedas, becomes the package for the existence within human frame. With this start with lesson, the initiation of Vedic systems in general and Vedic geometric systems in particular becomes complete.

■

8

GEOMETRY CLASS IN AYURVEDIC GARDEN

1. Initiations of young minds into Vedic systems is the most delicate sacred duty of the Vedic Mathematics teachers. And this duty is discharged by the teacher, the Vedic way.
2. The Vedic way of initiation of young minds into Vedic systems is to have initiation classes are held in Ayur Veda Garden where from Sathapatya layout to the whole range of transcendental phenomena of parallel transcendence Jyoti from rays of Sun manifesting as Ayur Veda Garden and Sama transcending either Veda plants, riding the Ati-Vahkas within rays of Sun and to reaching orb of the Sun, is fully blossoming and fragrance thereof spreading all round as ambrosia of bliss for the Sadhkas having an urge of initiations into Vedic systems.
3. The applied values initiations aspect is fulfilled by holding Vedic geometry based Vedic systems classes are held in Ayur Veda Garden and having interactive initiations with the growth

manifestations of whole ranges of organs of Ayur Vedic plants.

4. The second aspect of pure values of initiations is fulfilled by holding Vedic geometry based Vedic systems classes are held in Yaghya Shalas within Ayur Veda Gardens with focus upon the Sama ascendance phenomena putting into action by Agni (fire element); third element/third of Panch Mahabhuts.
5. The third aspect of transcendental values of initiations is fulfilled by holding Vedic Geometry based Vedic systems classes classes are held in Trinity of Gods temples within Yaghya Shala transcendental bases in Ayur Veda gardens.
6. The fourth aspect of Brahman-values initiation of Vedic geometry based Vedic systems classes are held focusing upon Being (self/Atman) and permitting the mind to transcend and attain self referral sustenance by continuous intensified glimpsing of the self referral sustenance of transcendental bases of temples of Trinity of Gods as Brahman source for the Brahman intelligence source of Brahman intelligence embedded into the Ati-Vahkas within rays of the Sun destined to carry Being for its onwards pilgrimage from within human frame to orb of the Sun.
7. The fifth and final aspect of Pra-Brahman values initiation of Vedic geometry based Vedic systems classes are held permitting the self referral sustenance of Being as on pilgrimage transcending the manifested frame, transcendental base, self referral sustenance and to be in unity state.

8. These different aspects of initiations, as different phases, sequentially focusing starting from Sathapatya lay out to growth of Ayur Veda plants, the emergence of Sama from Ayur Veda plants, the transcendental phenomena of Trinity of Gods, pilgrimage of Being from within human frame to orb of the Sun and even beyond to be unity state, are sequentially taken up parallel to unfolding of Sathapatya lay out of Ayur Veda Garden, growth of Ayur Vedic plants, transcendental phenomena of Trinity of Gods as source for transcendental bases of Yaghya Shalas, Being attaining breakthrough and riding Ativahkas within rays of the Sun and reaching orb of the Sun and then even transcending beyond and attaining unity state, are accordingly to be initiated in that sequence and order.

9. This, this way, as initiation phase of aspect of pure values within Yaghya shala in Ayur Veda Garden is the Aspect, about which the young minds are to be exposed about the fire as third of Panch Mahabhut (Earth, Water, Fire, Air, Space). The demonstrations of emergence of fire by forcefully rubbing wooden sticks, stones, and by focusing rays of the Sun may be of good help to bring home the physical manifestation of this element (fire). The fire flame rising upward is the another basic feature which deserves to be focused as the same is ultimately to be availed for pure value of ascendance concept of Vedic systems.

10. In this background, and like that, in sequential steps and with all delicate sensitivity for reaching applied values and physical manifested forms to

pure values and transcendental features of Vedic geometry based Vedic systems, deserve to be approached by the Vedic Mathematics teachers performing sacred duty of initiating young mind about the transcendental phenomena as Discipline of knowledge.

■

To know more

Jyoti (light) flows on Earth through rays of the Sun and Sama flows into orb of the Sun through rays of the Sun.

This transcendental phneomenon between Sun and Earth deserves to be chased. Within orb of the Sun is the lively existence domain (Vishnu lok /Atman / 6-space) of Divya Pursha.

Jyoti flow of this range is of six steps begining with Sun and reaching up till Earth. The reverse, begining with Earth and reaching up till Sun is the pathe of ascendance of Sama of equal range of Six steps but as of reverse orientation than that of the path of transcendence of jyoti flowing through the rays of the Sun.

This reversal of orientations is parallel to the shift from Domain fold to the Dimension fold as of the format of +1 space as domain accepting -1 space as dimension.

This as such would help us comprehend and appreciate the organisation of Vedic knowledge with Rig Ved as first Ved and Sam Ved as the third Ved of mantras of Rig Ved having been reset.

■

9

DIFFERENT PHASES AND STAGES

1. The Vedic geometry based Vedic systems' initiations of young minds within Ayur Veda gardens is of different phases and stages of transcendence and ascendance processes through different manifestation layers parallel to the organization of knowledge as Vedas.
2. The initiations as such deserve to be phased sequentially and the transition from one phase to another phase as well deserves to be properly cared by having suitable time gaps between two different phases.
3. The initial initiation of applied values in Ayur Veda garden but the first phase of initiation being about the Sathapatya layout of a garden in itself demonstrates well as that it is in a way a backward and forward moving process inherent in it as much as that at the very first step we are in Ayur Veda garden but the first phase of initiation is Sathapatya layout. This in a way going back from Ayur Veda garden to its Sathapatya layout and also having initiation into Sathapatya layout as first phase

initiation and reaching up till Ayur Veda is to have complete smooth transition from first phase to the next phase.

4. This way the initiation at each phase is to be planned and regulated in such steps that from the existing position there should in fact be a previous phase and following phase simultaneous exposures. The second phase of initiation exposure as about the growth range of different Ayur Vedic plants, naturally while being focused in Ayur Veda garden itself is to smoothly take to the layout of the Garden as well as to the manufacturing of Sama within the Ayur Veda plants.
5. The transition aspect of initiation should also take into account the sequential progress of transcending and ascending through one more manifestation layer at each phase of the initiation. The second phase of initiation about the growth ranges of Ayur Vedic plant, naturally takes through one additional manifestation layer stand that of the previous stage of Sathapatya layout format as manifestation layer. This, this way, while the initial first phase initiation if is to be of one manifestation layer then at the next phase this phenomena is to be of a pair of manifestation layers and at one step ahead at third phase of initiation, naturally the phenomena would be of three manifestation layers and like that it every step one additional manifestations layer would pile up and there by the transcendence and ascendance phenomena is to be along the whole range of manifestation layers.

6. Such being the phenomena, as that at every phase, additional manifestation layer is to emerge, which for the next phase is also to be transcended through and ascended there from, as such this aspect deserves to be constantly reminded and focused and properly taken care of, as otherwise, every manifestation layer may if not properly transited is to prove to be a mental block and the same if not properly taken care of preferable at initiation, then subsequently the ultimately it may turn out to be very prolonged process to melt away the initial mental blocks.
7. Here it also would be appropriate to mentioned from the experiences of senior Sadhka preserved in available Vedic scripture as that the Divya Pursha within orb of the Sun manifests the print out of Sathapatya Pursha on whose layout all phases of initiations may be chased. Pursha format to Divya Pursha with human frame in between and the two fold transcendence and ascendance through human frame with Pursha format of one side and Divya Pursha on the other side, is the three layered processing systems in terms of which along with a pair of another layers as upper most and lowest layers for the three layered processing systems would get potentialzed as five layered processing systems needed to process out the whole range of manifestations because of Panch Mahabhut.
8. As such the real attainment of transition is the transition from Panch Mahabhut manifestation to the sixth source manifestation as orb of the Sun.

9. Therefore, throughout initiation, the young mind have to be constantly reminded as that the transcendental phenomena of parallel transcendence of Jyoti through rays of the Sun and ascendance of Sama through rays of the Sun has to be through Panch Mahabhut manifestation layers which are to be chased starting with Sathapatya lay out Ayur Veda gardens till Sama reaching orb of the Sun.
10. The success of initiation is directly related to success of transition from the starting stage to the final stage of transcendental phenomena of parallel transcendence and ascendance through all the five manifestations layers because of Panch Mahabhut which are aimed to be crossed over as geometric steps of Vedic systems. And as such the sacred duty of Vedic Mathematics being to constantly remind the young minds about Sun as Atman/self and the Being is to have pilgrimage to orb of the Sun.

■

To know more

Pursha format is a 6-space format lively as Divya Pursha within orb of the Sun but when viewed from 3-space / Earth, the Pursha format is "human body format'. This chase be followed as beginning with cube and reaching up till cube end / cuboid with NVF (CUBOID) = 54 = NVF (SUN) = NVF (CAGED ONE).

■

10

CLASS IN A YAGHYA SHALA COMPOUND

1. For second phase initiation of Vedic geometry based Vedic systems, the shift is had from Sathapatya layout of Ayur Vedic Garden to Yaghyashala within Ayur Vedic Garden.
2. The Sathapatya way lay out of Ayur Veda Garden along Pursha format is organized to have Yaghyashala at center of the Ayur Veda Garden as Heart of Pursha format.
3. In fact, there is a Yaghyashala within each Ayurvedic plant as there is a Yaghyashala within human body and there being a Yaghya shala within orb of the Sun as well as within Brahmand and also within each ray of the Sun.
4. The Yaghyashala within Ayurvedic plants manufacture Sama; all Yaghya shalas within Ayur Vedic plants, human body, Brahmand, orb of the Sun and rays of the Sun are organized and coordinated as organization and coordination of whole range of Vedic knowledge as a single discipline.

5. The two fold transcendence and ascendance process of Jyoti and Sama make organization coordination of Yaghya Shala as spatial order organization of creator's Space with transcendental base because of which, with time the sacrificial fire (Agni Tatav in action in Yaghya shala) provides expanded expressions solid dimensional transcendental format for the manifested spatial order creations.

6. The young minds are to be initiated by taking the Sadhkas to Yaghya Shala with all delicate sensitivity of geometry based systems of Yajnas by sequentially exposing as to how the available Sama in manifested forms as different organs of Ayur Veda plants with the frequencies of Fire element provide expanded expressions format for spatial order of creator's Space as solid order of transcendental worlds.

7. Sequentially the young minds are to be exposed of transition from the manifestations format of hyper cube-4 to that of hyper cube-5.

8. It shall be focused as that hyper cube-4 gets multiplied as 10 hyper cubes-4 in 5-Space (transcendental worlds of Shiv Lok presided by Lord Shiv.

9. Then sequentially the young minds shall be exposed to single digit and double digit expressions for the numerals as:

Single digit	1	2	3	4	5	6	7	8	9
Double digit	01	02	03	04	05	06	07	08	09

10. Further young minds are to be exposed to the artifices of double digit numbers 04 and 04+05,

04+05+05, 04+05+05+05 and so on being 04, 09, 14, 19, 24, 29 and so on as a pair of flow lines with last digits as four and nine respectively. Here it would be relevant to focus that artifices of 4 and 9 are 2^2 and 3^2.

11. As such out of the ten flow lines, with last digits as, 1,2,3,4,5,6, 7,8,9, two of them as of last digits 4 and 9 are of such unique features where the spatial order (artifice of two) and solid order (artifice of three) as square units/second spatial order, take from manifested format to transcendental base, as much as that hyper cube-4 is enveloped within eight solids and center as seat of solid order accepts ninth solid.
12. With this the transcendence from manifested space (4-Space/domain of hyper cube-4) to transcendental worlds (5-Space/transcendental base of creator's Space) results into a provision for expanded expression for the manifested Sama.
13. This phenomena on artifices of numbers as 2^2 to 3^2 permits geometric chase as a plane of two facets and transcendence and ascendance between two such layers would amount to manifesting solid order amounting to synthesizing the pair of facets of a plane.
14. Further this with unfolding of the pair facets would amount to their reversal of manifestation. It is this manifestation of solid order and the manifestation of solid order into spatial order of double facet plane and single facet plane are the processes in plane being worked out and chased as Yajnas about which the whole initiation process

is to be had as second phase of initiation in Yaghya Shala within Ayur Veda garden.

■

To Know More

Here reaches a phase and stage where the focused attention is to be had as to how the ancient wisdom after emerging as a single system for whole range of knowledge organised as Vedas within Creator's space (4-space) had spread around.

The conceptual pillars like 'zero'. 'place value system', ' formats frames and forms' for notations of numerals and numbers are evidently of Vedic system.

The arithmetic to astronomy range, when chased from the original sources, as well woudl be convincingly acceptable as being of the same source. The 'fire altars' is one such original source whose chase is to enrich us about this range along with its conceptual pillars. The diagonal theorem of right angle triangle is preserved at base of the alphabet format itself coordinating 9 vowels, 25 varga (square format) consonants and 16 other consonants of quadruple groups of quadruple letters each (Anthstha, Ushmana and Yamas from Anuswara to Visergania, and four beyond).

From Samhita to Sulbha Sutra, there is a very big chain of scripture which are preserving the ancient wisdom.

■

11

MANIFESTATION FACETS

1. Young minds shall be initiated in Yaghya shala with all delicate sensitivity about manifestation of pair of facets of plane with the help of different organs of Ayur Veda plants.
2. Tulsi leafs, other Ayur Vedic leafs and other parts of Ayur Vedic plants may be availed with focus for this initiation about the pair of surfaces and parallel to it the geometric format of a plane with a pair of facets.
3. The content in between the pair of surfaces of leafs and the like being specifically focused for comprehension of geometric plane as well being such manifestation of a pair of facets with affine zero artifice geometric stuffs manifesting as such.
4. This manifestation be initiated as transcendence and ascendance from center of one surface to another surface of the leaf parallel to center of one facet of plane to another facet of the plane.
5. This transcendence and ascendance process also be chased as a three layered set up in two phases, first as transcendence from upper layer to the

middle layer and second as ascendance from lower layer to middle layer.

6. The transcendence from upper layer to the middle and ascendance from the lower layer to the middle shall be making the middle as manifestation as a distinct layer of pair of facets.
7. The reverse of the above process is to be as simultaneous ascendance and decadence from the middle towards the outer (upper and lower layers).
8. The middle layer with its ascendance and decadence is to make it a zero at artifice.
9. The double layered and single layered surfaces, synthesized, manifested, de-manifested and all processes simultaneously happening and all states simultaneously to be there, is all, to be focused, for proper initiation of the young minds to be sensitive about the Yajna of de-manifesting of manifested Sama parallel to the manifestation and de-manifestation processes of Yaghya Shalas within Ayur Vedic plants, human body, Brahmand, Orb of the Sun and rays of the Sun.
10. The focus upon the role of fire as third element for de-manifesting of manifested Sama is to be with all delicate sensitivity of the Yajna processes.

■

To know more

Idol of Creator avails four fold manifestation format for its manifestation. It is this hyper cube 4 format which is the source format to be chased for all manifestations including that of 'Sama'. being manufactured in the Ayur Veda gardens. ■

12

CONSTRUCTION OF FIRE ALTARS

1. Yaghyashala is basic Vedic institution. It is Vedic laboratory.
2. The central part of this laboratory is Vedi (Altar).
3. The construction of Vedi deserves to be attended with geometric precision.
4. This discipline at the apex accepts Yajur Ved as a source.
5. Further this discipline, at the apex level of Up vedas, takes to the lineage of Sathapatya Up Veda.
6. As Vedanga, it is part of Kalpa-Vedanga, which is one of the six Vedangas (Shiksha, Vyakarna, Nirukat, Chandas, Jyotish and Kalpa). This as an discipline of construction of Vedis, that way is part of Kalpa Sutras.
7. This as independent discipline, as construction of Vedis, is also designated as Sulbha Sutras.
8. Yajnas, as applied values, predominate the Brahmanas (Shatpath Brahamna and other Brahmanas).

9. As Darshna, the Yajna institution, is more prominently, availed as Mimansa Darshan.
10. The basic geometrical concept availed by the discipline of construction of Vedis is that of MEASURE and the basic measuring instrument being availed is Yaju (Rope/measuring tape).

INITIATION OF FIRST SKILL TO MEASURE HALF UNIT:

Conceptual format:

One (1) as double digit (01) with its reflection pair as 10, sets geometric initiation with the help of Yaju (as of one unit, to be viewed as of ten units), to have half unit length by bringing both ends (01 as well as 10) along one side and the middle to have at half length of the rope, as (half, .5, and as 5).

Practical steps:

1. To have a rope.
2. To accept the whole length as of one unit by one unit (01).
3. To accept this unit as (10).
4. To bring both ends of the rope on one side.
5. Fix both ends of rope in one hand and starch both strings of the rope and to hold the joint in another hand.
6. The starched length of pair of strings of rope between pair of hands is of half length of the rope. It is taken as half unit length. It is also accepted as half of ten that is 5.

Exercise:

1. To fix middle point between pair of Ayur Vedic

plants.

2. To measure half distance between a pair of pegs fixed in ground for construction of Vedi.
3. To mark middle points of units on a scale.
4. To construct a scale of units of 1 1/2 times than that of units of given scale.

■

To know more

In fact the learning of applied values of geometric formats begins here in the construction of the fire altars.

The caution here is that the over emphases upon the applied values translation into the mundane properties parallel to the linear order set ups of macro state is to cost more as the same in the process is to result into 'bios' which shall be resulting into mental blocks sealing every thing only up till the sensory field and all beyond which is there as intelligence field and even beyond, the consciousness field would never come lively for such minds.

Therefore, heavy responsibility lies upon the teachers of Vedic geometry to remain ever conscious to this crucial junction to ensure as that all mental blocks are to be simultaneously to be countered and to be melted for the desired transition from the macro state linear order to the micro state spatial order and beyond that to the transcendental state of solid order.

■

13

VALUES OF VEDIC KNOWLEDGE AND YAJNAS

1. Pure values and applied values of Vedic knowledge, are two different sets of values in terms of which the Vedic knowledge may be approached as pure knowledge and applied knowledge.
2. Though pure values and applied values run parallel to each other but while pure values run parallel to the transcendence through the manifestations, the applied values run parallel to the ascendance.
3. The ascendance of Sama and transcendence of Jyoti, as pair of parallel flow lines get categorized as applied values and pure values on the basis of their start with sources; Sama, being manufacture here in medicinal plants, as such in its ascendance yields the applied values and on the other hand Jyoti with its source as orb of the Sun, its flow as transcendence makes its flow as pure values.
4. Parallel to the transcendence of Jyoti flow as Shrutis, is cognized as Vedas as the pure values. However parallel to ascendance of Sama flow as

samritis, is recognized as Vedas as applied values.

5. Accordingly, pure values transcend parallel to Jyoti flow as fruits of Yajnas in the cosmos and applied values ascend parallel to Sama flow as fruits of Yajnas here in Ayur Veda Gardens.

6. The initiations of young minds in Yaghshalas in Ayur Vedic gardens, as such, essentially being the applied values parallel to ascendance of Sama, therefore heavy responsibilities lies upon the Vedic Mathematics teachers to take every care that the young minds do not get any block which may in any way come in the way of cognition of the pure values parallel to transcendence of Jyoti through rays of the Sun.

7. Full focus be had upon the transition aspect from this phase of initiation in Yaghyashala to the following phases of pilgrimage of Being to orb of the Sun and even beyond.

8. For this, it would be appropriate that at appropriate intervals during this phase of initiations, the Sadhkas shall be encouraged to go intrans frequently and to permit the mind to transcendence to glimpse the inner folds of the transcendental world time and again and to get fulfilled with ambrosia of bliss of the inner folds of the transcendental world.

9. Basic approach to Vedic knowledge and its systems is to be of self referral sustenance for the transcending mind and as such the stage of chasing during trans of the transcending mind is to be attained for perfection of intelligence and it

is only with perfection of intelligence, it may be possible to have proper initiations.

10. The chase of the Yajnas processing of Sama ascendance with perfected intelligence is to be of the order of the parallel phenomena of cognizance of Shrutis parallel to the transcendence of Jyoti from orb of the Sun along its rays manifesting as Ayur Veda gardens.

■

To Know More

Sadkhas shall perfect their intelligence to glimpse the Yajna continously being performed within their body.

One is to be conscious for the steps to ensure a smooth transition from the sensory field to that of the intelligence field, and a step ahead, to that of the consciousness field for being carried along the transcendental carriers within the rays of the Sun to the orb of the Sun, and even beyond that to the Sapt Rishi lok.

One is to be of blissful attainment for the take off of the Being from the sixth eternal circuit of the human frame and to ride the rays of the Sun.

One is further to be blissful with glimpsing of the inner folds and inner most core of the transcendental worlds for the previledge to be carried by the Brahman intelligence embedded transcendental carriers to the Brahman domain, the ultimate destination.

■

14

ARITHMETIC TO ASTRONOMY RANGE

1. The success of initiations of young minds in the Yaghyashala within Ayur Veda garden may be measure firstly in terms of the perfection of intelligence and secondly in terms of the yajnas range to be parallel to the Arithmetic to Astronomy range of applied values of Vedic knowledge.
2. For the first goal, the Sadhkas are to be encouraged to go intrans time and again and to glimpse the inner folds of transcendental worlds.
3. However for the second goal, the initiations are to begin with the geometry of constructions of Vedis but the discipline of geometry is to go deep to the formats of manifestations and ascendance there from of the Sama and the whole process is to carry along with it the whole range of processing in terms of Arithmetic to astronomy range of applied values of Vedic knowledge.
4. The full exposure of young minds at the time of their initiation in Yaghya shala be attained by bringing the library within the Yaghyashala and scripture after scripture be physically be taken

out and relevant representative portions be recited to bring home the point as that whole range of knowledge runs parallel to the Yajnas approach to reality.

5. Starting with Yajur Veda and reaching up till the Sulbha sutras and then starting with Sulbha sutras and reaching up till Yajur Veda, whole range is to be chased both phase beginning to end and reverse thereof.
6. The major Sulbha sutras of Baudhayana, Apastanmba, Manava, Mairayana and Varaha belonging to Krishna Yajurveda and other Sulbha sutras belonging to other Samhitas, be taken up one by one and the same be introduced. Baudhayana sulbha sutra is organized as three chapters of 116 Sutras, 86 Sutras and 323 Sutras with internal organizations taking up sequentially the different geometrical aspects of constructions of Vedis. These organizations be introduced chapter wise and chapter be introduced sequentially, the way the aspects are dealt with in this Sulbha Sutra.
7. Likewise the other sulbha sutras be taken up one by one and same be introduced in the sequence and order of organization of knowledge and skills in these sulbha sutras.
8. The initiations shall aim to have full exposure in terms of the available Vedic literature.
9. Initiations shall also aim to have full exposure in terms of the experiences of senior Sadhkas continuing there Sadhna through the medium of Sadhnas.

10. Initiations shall further aim to have full exposure of the way Sama is being manufactured by different Ayur Vedic plants and the way Sama ascends through Ayur Veda gardens and reaches the orb of the Sun.

■

To Know More

Arithmetic to astronomy is the intellectual range. beyond that is the transcendental range which is even beyond the knowledge as at that phase and stage, the knowledge itself goes transcendental and everything emerges to be the transcendental phenomenon and it transcendentally unfolds and folds back of its own.

■

A words with the teachers about middle school mathematics topics

A WORD WITH THE TEACHERS

Respected teachers,

With all humbleness, it is being shared as that the realistic approach for initiating young minds in the Discipline of mathematics, is to take them along the format of a **CUBE (A)**, as

__ The all **'WORLD'** around is a manifested domain, and

__ The '***MIND***' as well is seated within a framed body;

And, as such, **everything** with which senses as first **PROCESSOR** of **MIND** are to interact with and to revert back from their for creating a processing domain for the mind, are **solids**. ■

The points, lines and surfaces of such solids are nothing but solid points, solid lines and solid surfaces.

This reality of first reach, as it is would be soothing enough for the young minds to be comfortable for handling them as such, as solids.

Moreover, the topics of mathematics, being covered by middle school mathematics are of formats of lines, planes and solids as an axis, a pair of axes and a set of all the three axes of 3-space being at work.

As such these three set ups of points, lines and surfaces are to be taken up of all the three axes to be of zero value for the points, a pair of axes to be of zero value for edges and one of the axis being of zero value for the surfaces of the cube. However in case of the solids / domain of cube, all the three axes are to be taken of positive values.

Cube, as such is to be taken as a representative regular body of 3-space as its whole set up, as such, may it be its corner points, edges or surfaces or domain as being solid – content lump as coming within the comprehension range of 'senses'.

Further as is going to be self evident from the following chase of specific topics of this phase and stage of schooling for the discipline of mathematics up till middle level, stage as that, this feature and comprehension of the set up of the cube is only a specific feature of this set up, which otherwise for its generality, is to be approached along the manifestation format of Creator's space (4-space).

In this background, here the class wise topics are being glanced to have to have their appreciation at conceptual formats.

CLASS-1

1. The learning topics of this phase and stage are (1) Numbers 1 to 100 (2) Addition and subtraction, and (3) Money, time and measurement as practical illustrations of applied values of learning.
2. These topics and range of learning is to take help of 'counts' and their ordering principle of 'one more'.
3. Once the counting pebbles and abacus beads are replaced by 'cubes' and to 'functional format of Ganita Sutra-1:

   'Ekadhiken Purvena' / One more than before' is availed, the whole 'transition for switch over to the firm foundation of Vedic mathematics would be attained without any sacrifice of current teaching and the desired argumentation of conceptual base for shift to the Vedic mathematics systems would be following of its own.
4. 'Cube' being a structured set up of the order of the manifestation format of Creator's space, it shall be providing us not only the conceptual basis but also the functional base for approaching numbers along with geometric formats.
5. The grid frames acceptable to the surfaces and solid domain accepting cages as of solid grids shall be ensuring wonder for the coordination of artifices of numbers parallel to dimensional frames.
6. So approached, the numbers would unfold their all features transcending every limitation of

manifested restriction. It is in terms of it, that the multiplication operation would be free from its initial emergence as only a repetitive addition. And ahead, likewise, the division and exponents as well, would get freedom from the restrictive play as repetitive subtraction as division and as repetitive multiplication being exponents.

7. Still further, numbers themselves would get freedom from their single digit formats being '1, 2, 3, 4, 5, 6, 7, 8, 9) and would be of their blissful roles being '01, 02, 03, 04, 05, 06, 07, 08, 09', and, then as '001, 002, 003, 004, 005, 006, 007, 008, 009', and ahead of as wider ranges, as can be the range of counts themselves.
8. With it '—00001' itself would be the whole range of counts, and at the same time remaining fully expressed at first place of any place value system.
9. It is this 'conceptual format' for which the foundation deserves to be laid at the initial stage, otherwise at subsequent stages, it would be too late to expect desired correction for the manifested mental blocks emerging at every such conceptual comprehension steps.

The needed transition from repetitive addition to 'multiplication', is to be acquired by acquiring skill of handling cross format ment for spatial order wo digits formats for numbers including numerals, at unit place of the place value system as 'requiring summation' of the pairings, while the same for 'two' and more digits format would be requiring instead of summation for the pairings results, same to be taken along with their place values. The illustration of (1 , 1) as pairing for unit place is 1+1=2 (two) while the same for two places format is to be 11 (eleven).

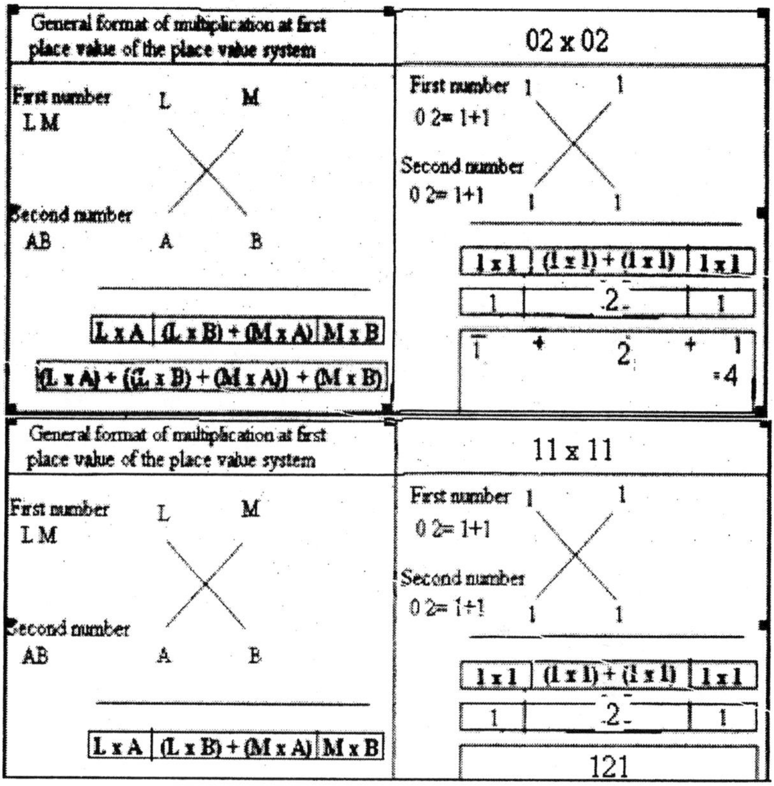

1. This, as such would, accordingly, yield multiplication at unit place and at pair of places, while availing the same cross format as depicted above.
2. And, it is this conceptual format comprehension, which need be attained as but for it, the whole range of structural features would get evaporated and with it one would be deprived off 'intelligence, knowledge as well as the wisdom'.

CLASS-2 TO 4 AND CLASS-5 TO 7

1. The learning process of the middle stage schooling, as per the choice of topics of mathematics, makes it of two phases, of which the first takes from class 1 to class 4 and the second phase takes from class 5 to class-8.
2. The topics of class-1 to class-4 accept grouping in terms of approaching them with a focus upon 'whole numbers', while the topics of class-5 to 8 accepting grouping in terms of approach to them with a focus upon 'fractions'.
3. The approach as whole numbers and as fractions / rational numbers have very fine distinctions which deserve to be attended to with equal sensitivity of the delegacy.
4. The topics of class-2 to 4 are as follows

   2.1 Numbers 2.2 Addition and subtraction 2.3 Multiplication 2.4 Geometrical shapes 2.5 Measurements

   3.1 Numbers 3.2 Addition 3.3 Multiplication 3.4 Geometrical shapes 3.5 Time 3.6 Subtraction 3.7 Length, Mass (or weight) and capacity 3.8 Division 3.9 Money 3.10 Fractional numbers 3.11 Pictorial representation of data

   4.1 Regional and Roman Numbers 4.2 Numbers 4.3 Factors and Multiples 4.4 Fractional Numbers 4.5 Pattern **THE FOUR BASIC OPERATIONS** 4.6 Addition and Subtraction 4.7 Multiplication 4.8 Division 4.9 Pictorial Representation of Data **MEASUREMENT** 4.10 Length 4.11 Mass (or Weight) 4.12 Capacity 4.13 Money 4.14 Time 4.15 Geometrical Shapes

5. The topics of class-1 to 4, accept grouping as (A) numbers (B) Four basic operations and geometric shapes (D) Applied values of measurements and other connected situations.
6. The numbers' basic operations, can be skillfully handle with proficiency, efficiency and speed by availing geometric formats supplied by the set up of the cube in the forms of corner points, edges, surfaces and domain. This set up, as it is, provides us lines with slots (points), surfaces with grid and domain with solid grids, and the same are to be availed for concrete formats for the facility of the young minds.
7. Perfection of the skill of counting, direct as well as reverse, is the first step of learning. For it the format of the edges with a pair of orientations would be best suited for direct and reverse counting. The edge itself is to be accepted as a track of a moving point shifting step by step from slot to slot as a count followed by a count.
8. Then, the young minds can be smoothly initiated for transition from the format of a line with slots to a pair of lines with slots and corresponding slots within the plane as grid zones. This attainment would be a very big step of learning. With this attainment, the learning would really achieve a shift from a line format to that of a grid format.
9. The grid format is a format of a pair of axes / edges. It is a format of spatial order. Its shifts us from the boundary of a square to the domain of a square. It takes us within the domain of a plane / grid zone.

10. The grids of as many rows and column can be constructed. It is this activity of constructing grids of given number of rows and columns which is to achieve all the wonders of approaching the numbers along a place value system. It shall be also providing the formats for expressing big numbers in terms of only ten symbols / numerals 1 to 9 and the 10th place value symbol / '0'.

11. The topics of class-5 to 7 are as follows

    5.1 Numbers 5.2 Basic Operations on Numbers 5.3 Highest Common Factor and Lowest Common Multiple 5.4 Fractional Numbers 5.5 Basic Operations on Fractional Numbers 5.6 Decimals 5.7 Basic Operations on Decimals 5.8 Measurement 5.9 Money Transaction 5.10 Pictorial Representation of Data 5.11 Basic Geometrical Concepts 5.12 Triangles 5.13 Circles 5.14 Perimeter, Area and Volume Answers

    6.1 Natural Numbers and Whole Numbers 6.2 Integers 6.3 Factors and Multiples 6.4 Ratio, Proportion and Unitary Method 6.5 Percentage and its applications 6.6 Algebraic Expression 6.7 Linear Equation in One Variable 6.8 Basic Geometrical Concepts 6.9 Line Segments 6.10 Angles 6.11 Pairs of Lines and Transversal 6.12 Triangles 6.13 Constructions 6.14 Perimeter and Area

    7.1 Rational Numbers 7.2 Operations on rational numbers 7.3 Decimal representation of rational numbers 7.4 Exponents 7.5 Direct and inverse variations 7.6 Percentage and its applications 7.7 Algebraic expressions 7.8 Factorization of algebraic expression 7.9 Linear equation in one variable

7.10 More about triangles 7.11 Congruent triangles 7.12 Quadrilaterals 7.13 Circles 7.14 Areas of rectangular paths 7.15 Surface areas and volumes 7.16 Statistics

CLASS-8

The topics mentioned are 8.1 Square and Square roots 8.2 Cubes and Cube Roots 8.3 Rational Exponents and Radicals 8.4 Profit, Loss and Discount 8.5 Compound Interest 8.6 Algebraic Identities 8.7 Polynomials 8.8 Equations in One Variable 8.9 Parallel Lines 8.10 Special Types of Quadrilateral 8.11 Construction of Quadrilateral 8.12 Circles 8.13 Areas 8.14 Surface Areas 8.15 Volumes and 8.16 Statistics.

12. The Discipline approach of branching as (A) Arithmetic, (B) Algebra (C) Geometry and (D) Applications, here is on its reverse, that is, a unification expression, as that all these four folds feature are taking us back to the situations where these are unified in their expressions. Illustratively square an square roots, are arithmetic situations, algebraic situations, geometric situations, as well as the applied values situations.

13. Still the topics subject matter of instructions, as these are, are being re-grouped, in terms of their prominent features namely, under

    Group 1, as arithmetic group as 8.1 Square and Square roots 8.2 Cubes and Cube Roots 8.3 Rational Exponents and Radicals.

    Group 2, as algebraic group as 8.6 Algebraic

Identities 8.7 Polynomials 8.8 Equations in One Variable.

Group 3, as geometry group as 8.9 Parallel Lines 8.10 Special Types of Quadrilateral 8.11 Construction of Quadrilateral 8.12 Circles 8.13 Areas 8.14 Surface Areas 8.15 Volumes, and

Group 4, as applications group, as 8.4 Profit, Loss and Discount 8.5 Compound Interest 8.16 Statistics.

TO BE APPROACHED AVAILING SET UP OF A CUBE

1. The sequential flow through edges, and edges being the tracks of moving points shall be providing us the needed basis for equations in one variable which is to be availed as geometric format for the values of the variable.
2. The edges constituting boundary of surface plates are to be availed for the 'parallel lines', and for the computation of areas (including surface areas), and for construction of quadrilateral.
3. The centers of the 'surface plates' of the cube, are to be availed as centers of the circles, and concentric circles may be constructed and worked out.
4. The squares and square roots can be approached geometrically by availing squares set ups of the surface plates of the cube.
5. The grids frame for the surface plates of the cube shall be supplying the necessary transition from geometric formats to arithmetic formats for approaching squares and square roots in terms of artifices values of numbers.

6. The features of rational exponents and radicals as well can be approached in terms of the grids frame of the surface plates of the cubes as well as for the solid grid cages for the domain of the cube.
7. A shift from the 'spatial grid frames' of surface plates of the cube to solid grid frames of the domain of the cube, is to be taken as a transition step from that of the shift from the slots (points) along the edges to the spatial grid along the surface plates.
8. The solid grids cages for the domain of cube shall be providing solid units for computation of the volume of the cube, and also for the solids to be approached as set up of appropriately selected unit cubes.
9. The format for representation of data with frequencies is to be supplied by the spatial and solid grids.
10. The transition from spatial grids to the solid grids is to emerge as the take off stage for the next phase and stage of leaning.

CONCEPTUAL ATTAINMENT

1. Focus is to be upon the conceptual attainment of this phase and stage of learning.
2. The basic conceptual formats of this phase and stage of learning are of (1) One variable (2) Parallel lines (3) Square (4) Quadrilaterals (5) Circle (6) Surface areas (7) Cube (8) Volume (9) Rational exponents and radicals (10) Polynomials.
3. The functional approach to these conceptual set ups is of (1) 'Edge' as a moving point track and flow

stream (2) 'Surface – plate' as a moving edge manifesting parallel lines track accepting grid frame, and (3 'volume', as a moving surface plate manifesting parallel surface plates track accepting a solid grids as cages for domain of the cube).

4. This sequential chase, as rational exponents and radicals, is to be taken as a phase and stage of transition for the further take off to the set up of hyper cube 4.

SKILLS FOR THE APPLIED VALUES

1. The beginning to be had with comprehension of the features of 'one' as a unit acceptable to the set ups in terms of their geometric bases.
2. The unit for 'edges' as lines is to be of different features than that of units for surface plates and domain of the cube.
3. The generation status of edges as tracks of moving corner points, surfaces as tracks of moving edges creating parallel lines (and grids), and solid domain of the cube as a track of moving surfaces, inherently go into the setups of the units of lines, surfaces and domains as per their generation status, namely first degree / power unit (1^1) for lines / edges, second degree / power unit (1^2) for surfaces and third degrees / power unit (1^3) for solid domain of the cube.
4. The radical exponents as these generation status power units are, viz., (1^1, 1^2, 1^3), these accept extension to 1^4, 1^5 and so on, but these remain simply as 'repeated exponents', and the same puts these repeated exponents values parallel to the repeated addition and repeated multiplication, to

be not of 'full expressions' and further like 'repeated addition' which needs the spatial format (and of double digit numerals) for full expression for the multiplication and ahead for freedom of the exponents from the restriction of 'repeated multiplication features', there is a need for solid format and of triple digit numerals. So, for the 'transcendental units', to be free from the restrictions of 'repeated exponents', there would be a need for transition and a shift from the format of a cube to that of 'hyper cube 4'.

5. This being the conceptual need, as such, the set up of the cube itself deserves to be approached afresh from that of its linear order of 3 linear axis, to that of the four fold manifestation along the manifestation format of Creator's space (4-space).
6. It is only by such approach to the set up of the cube as being four folds, namely, as that 1-space is to play the role of dimension order of this setup, 2-space to be in the role of boundary fold, 3-space content lump is to play the role of domain fold, and finally the 4-space is to play the role of origin fold of this set up, that the necessary conceptual breakthrough can be attaining, failing which the reality comprehension is to be held up only up till the geometric envelope of the cube, that is up till spatial boundary of 2-space, and the 'domain of the cube' is to be every elusive.
7. This being so, heavy responsibilities lies upon the teachers to expose the young minds to the set up of cube, as availing the four fold manifestation format of 4-space, with center of the cube being

the seat of hyper cube 4 / 4-space / Creator's space.

8. It is only with comprehension of the structural set up of the center of the cube as seat of hyper cube 4 that it may be taken as that appropriate conceptual foundation for perfection of intelligence of the young minds has been laid down.
9. It is only with comprehension of the structural set up of the cube, the young minds shall be laying down spatial order for their mental formats.
10. It is only with spatial order for the mental format that the pairing operation would be within the mental reach.
11. It is only with the pairing operation as that, it my be appreciated as the functional format of linear operations to be restricted for their approach in terms of dimensional frames of half dimensions.
12. The proficiency of skills is possible only in terms of the degree of efficiency with which the computations of geometric formats are worked out in terms of dimensional frame of half dimensions.
13. This, in fact amounts to attaining double units along second axis while the working is to be in terms of half units along the first axis.
14. It is this wonderful feature of 'taking half steps' and attaining 'two steps', which is being en-cashed by the Vedic systems.
15. One may see that ' ½ to 2' is '1 to 4'. The sole syllable Om / Ek Akshar Braham, manifests four padas. 'One Ved' gets organized as 'Four Vedas. Interval as domain, in its role as boundary of 2-

space gets multiplied four folds as much as that the domain boundary ratio for 2-space is A^2: 4 A^1 , which is there as the boundary of 2-space / square is of 4 linear components.

16. It is this spatial order features, which stitches synthetic boundary for a cube as of 26 entities (8 points, 12 edges, 6 surfaces). It is this order which takes to half dimension / half edge as a functional format. It brings into a reflection pairing of halves, which as artifices pairs (01, 10). The shift to ten place value system, as such amounts to working in terms of 'half dimensions' and also amounts to availing 'reflection paring features'. This is at the base of their being precisely 26 primes up till 10 x 10 = 100. These primes by definition as accepting no divisor other than one and itself, brings one itself as a prime. The range 26 primes comes to be 1, 2, 3, 5, 7, 11, 13, 17,19, 23, 29, 31, 37, 41, 43, 47, 53, 59, 61, 67, 71, 73, 79, 83, 89, and 97.
17. Now if the unit for an edge be shifted from 01 to 10, It would split cube as 1000 cubes. As within 4-space, 'zero' and 'unit' are of same number value format, as such the two parts of the dimension as '0' and '1' shall be constituting a set up for three dimensional frame as eight triples, namely, (0, 0, 0), (0, 0, 1), (0, 1, 0), (1, 0, 0), (0, 1, 1), (1, 0, 1), (1, 1, 0), and (1, 1, 1). The sum of these triples values yields 0, 1, 1, 1, 2, 2, 2, 3. The frequency arrangement of these values would mean zero value occurs once while values 1 and 2, each to occur thrice and the value '3' to occur only once.
18. This leads to following tabulations

Value	0	1	2	3
Frequency	1	3	3	1

19. Now the units flow from either side of the edge shall be yielding double unit at the middle. It amounts to a set up of the order (1, 1+1, 1), that is (1, 2, 1). Such format being there for all the three axes, the same together shall be working out (3 x 1, 3 x 2, 3 x 1), that is (3, 6, 3). This as along a linear format is to be of artifice 363 and the same with '10' as unit for the linear format, shall be yielding 3630 linear units set up.

20. The three dimensional frame shall be accepting split as a pair of three dimensional frames of half dimensions. This split as such as a pair of three dimensional frames of half dimensions along an axis, and then there being a pair of axis for the surfaces shall be making a double pair of three dimensional of half dimensions and pair and these pairs while synthesizing within spatial order, shall be yielding double pair of three dimensional frames of half dimensions into a pair of three dimensional frames of full dimensions, and there would be available a pairing accepting expression as artifices as (3, 3). This as a chase for the spatial format 10 × 10 = 100, and a step ahead (3, 3) as 33, as such 3300 components of spatial order units.

21. The points format leading 1331 units, the edges format leading to 3630 units and surfaces format leading to 3300 units, when emerge to be precisely the number of corners, edges and surface plates of 1000 cubes, it sends a strong geometric message of Reality to be comprehended while chasing the

set up of a cube, as a manifestation along the manifestation format of a Creator's space (4-space).

It would be a blissful exercise to chase the above again to have respectful understanding with the **cube** as our first friend from the world of creations.

SUM UP AND RECAPITULATION

1. 'Cube' is one word in terms of which the learning of this phase and stage may be summed up.
2. 'Cube' is the set up which is the answer for all 'teaching and learning' situations through which one is to be through.
3. 'Cube' is the format, as well as the manifestation, and is answer for its package as its linear order and the same is also the poser for its needed transition to the spatial order reality ahead.
4. 'Cube' is solid, as well as, it is hollow. It is enveloped within a spatial boundary, as well as it envelops the hyper solid domains.
5. 'Cube' is a linear order set up and works out solid order for the transcendental worlds, as a transcendental phenomenon of a transcendental feature of pairing leading to NVF (LINEAR) = NVF (SOLID).
6. 'Cube' at end of the range of artifice 26 emerges as 'cuboid' being the 'air god' of the order of the 'Sun'; NVF (SUN) = 54= NVF (AIR GOD); air is the fourth element while Sun is the sixth element and these accept pairing as 4-space playing the role of dimension of 6-space as Lord Brahma, Creator the Supreme and presiding deity of 4-space

is also the presiding deity of the measure while Lord Vishnu is the presiding deity of 6-space, as well as of the measuring rod.

7. 'Cube to cuboid' is the transcendental manifestation range as 'cube' / 3-space as solid dimensional order of transcendental worlds (5-space) has Vishnu lok (6-space) as its origin source.
8. The cube manifests spatial boundary as much as that its corners points (0-space entities) play the role of dimension, edges (1-space entities) play the role of boundary, surfaces (2-space entities) express as domains and solid domain (3-space) set up emerges as the origin source.
9. Cube within synthetic boundary with corners as points playing the role of origin for the spatial boundary, deserve to be chased further, as that these are eight in number and as each of it is equipped with a 3 dimensional frame of half dimensions with and the orientations for the axis of these dimensional frames being inward.
10. Cube with these features of the set up of its corner points reflect of the for the split as well as the synthesis of the spatial boundary enveloping wihtin a solid domain as the source reservoir for sustaining flow streams at the boundary.
11. Cube as a solid domain with creator's space as its core base and a source reservoir feeds the solid content flow streams which ultimately manifest solid boundary for the creator's space.
12. Cube as a linear order set up manifests this order as edges at the boundary and these though are twelve in number but only seven of them are

sufficient to carry the flow stream through all the eight corner points.

13. Cube is the set up which sustains every flow stream through all the eight corner points in terms of only seven of its edges, to be designated as the manifested edges while the remaining five edges in the context being unmanifest edges have a distinct role to play of providing un manifest base for the manifested set of edges of the cube.

■

Other Books on

VEDIC MATHEMATICS SERIES

1. Vedic Mathematics New Horizons
   Initial Lessons (New)
2. Vedic Mathematics New Horizons
   Advance Lessons (New)
3. Vedic Mathematics for All
4. Learn and Teach Vedic Mathematics
5. Practice Vedic Mathematics
6. Vedic Mathematics Decodes Space Book
7. Vedic Mathematics Basics
8. Vedic Mathematics Skills
9. The Teaching of Vedic Mathematics
10. Learning Vedic Mathematics on First Principles
11. Vedic Geometry Course

Lotus PRESS

Unit No. 220, Second Floor, 4735/22,
Prakash Deep Building, Ansari Road, Darya Ganj,
New Delhi - 110002, Ph.: 41325510, 23280047, 09811838000
E-mail: lotuspress1984@gmail.com, www.lotuspress.co.in